MORRIS MINOR
Collection No.1

Compiled by
R.M. Clarke

ISBN 0 907 073 522

Distributed by
Brooklands Book Distribution Ltd.
'Holmerise', Seven Hills Road,
Cobham, Surrey, England

Brooklands Books Titles in this series

AC Cobra 1962-1969
Armstrong Siddeley Cars 1945-1960
Austin 7 in the Thirties
Austin Seven Cars 1930-1935
Austin 10 1932-1939
Austin Healey 100 1952-1959
Austin Healey 3000 1959-1967
Austin Healey 100 & 3000 Collection No. 1
Austin Healey Sprite 1958-1971
Bentley Cars 1919-1929
Bentley Cars 1929-1934
Bentley Cars 1934-1939
Bentley Cars 1940-1945
Bentley Cars 1945-1950
BMW 1600 Collection No. 1
BMW 2002 Collection No. 1
Buick Cars 1929-1939
Camaro 1966-1970
Chrysler Cars 1930-1939
Citroen Traction Avant 1934-1957
Citroen 2CV 1949-1982
Corvette Cars 1955-1964
Datsun 240z & 260z 1970-1977
De Tomaso Collection No. 1
Dodge Cars 1924-1938
Ferrari Cars 1946-1956
Ferrari Cars 1957-1962
Ferrari Cars 1962-1966
Ferrari Cars 1966-1969
Ferrari Cars 1969-1973
Ferrari Cars 1973-1977
Ferrari Cars 1977-1981
Ferrari Collection No. 1
Fiat X1/9 1972-1980
Ford GT40 1964-1978
Ford Mustang 1964-1967
Ford Mustang 1967-1973
Ford RS Escort 1968-1980
Hudson & Railton Cars 1936-1940
Jaguar (& S.S) Cars 1931-1937
Jaguar (& S.S) Cars 1937-1947
Jaguar Cars 1948-1951
Jaguar Cars 1951-1953
Jaguar Cars 1954-1955
Jaguar Cars 1955-1957
Jaguar Cars 1957-1961
Jaguar Cars 1961-1964
Jaguar Cars 1964-1968
Jaguar Sports Cars 1957-1960
Jaguar E-Type 1961-1966
Jaguar E-Type 1966-1971
Jaguar E-Type 1971-1975
Jaguar XKE Collection No. 1
Jaguar XJ6 1968-1972
Jaguar XJ6 1973-1980
Jaguar XJ12 1972-1980
Jaguar XJS 1975-1980
Jensen Cars 1946-1967
Jensen Cars 1967-1979
Jensen Interceptor 1966-1976
Jensen-Healey 1972-1976
Lamborghini Cars 1964-1970
Lamborghini Cars 1970-1975
Land Rover 1948-1973
Lotus Elan 1962-1973
Lotus Elan Collection No. 1
Lotus Elite & Eclat 1975-1981
Lotus Esprit 1974-1981
Lotus Europa 1966-1975
Lotus Europa Collection No. 1
Lotus Seven 1957-1980
Lotus Seven Collection No. 1
Maserati 1965-1970
Maserati 1970-1975
Mazda RX-7 Collection No. 1
Mercedes Benz Cars 1949-1954
Mercedes Benz Cars 1954-1957
Mercedes Benz Cars 1957-1961
Mercedes Benz Competition Cars 1950-1957
MG Cars in the Thirties
MG Cars 1929-1934
MG Cars 1935-1940
MG Cars 1940-1947
MG Cars 1948-1951
MG Cars 1952-1954
MG Cars 1955-1957
MG Cars 1957-1959
MG Cars 1959-1962
MG Midget 1961-1979
MG MGA 1955-1962
MG MGB 1962-1970
MG MGB 1970-1980
MG MGB GT 1965-1980
Mini-Cooper 1961-1971
Morgan 3-Wheeler 1930-1952
Morgan Cars 1936-1960
Morgan Cars 1960-1970
Morgan Cars 1969-1979
Morris Minor 1949-1970
Morris Minor Collection No. 1
Nash Metropolitan 1954-1961
Packard Cars 1920-1942
Pantera 1970-1973
Pantera & Mangusta 1969-1974
Pontiac GTO 1964-1970
Pontiac Firebird 1967-1973
Porsche Cars 1952-1956
Porsche Cars 1957-1960
Porsche Cars 1960-1964
Porsche Cars 1964-1968
Porsche Cars 1968-1972
Porsche Cars 1972-1975
Porsche Cars in the Sixties
Porsche 914 1969-1975
Porsche 911 Collection No. 1
Porsche 911 Collection No. 2
Porsche 924 1975-1981
Porsche 928 Collection No. 1
Porsche Turbo Collection No. 1
Riley Cars 1932-1935
Riley Cars 1936-1939
Riley Cars 1940-1945
Riley Cars 1945-1950
Riley Cars 1950-1955
Rolls Royce Cars 1930-1935
Rolls Royce Cars 1935-1940
Rolls Royce Cars 1940-1950
Rollys-Royce Silver Shadow 1965-1980
Range Rover 1970-1981
Rover P4 1949-1959
Rover P4 1955-1964
Singer Sports Cars 1933-1954
Studebaker Cars 1923-1939
Sunbeam Alpine & Tiger 1959-1967
Triumph Spitfire 1962-1980
Triumph Stag 1970-1980
Triumph TR2 & TR3 1952-1960
Triumph TR4 & TR5 1961-1969
Triumph GT6 1966-1974
TVR 1960-1980
Volkswagen Cars 1936-1956
VW Beetle 1956-1977
VW Karmann Ghia Collection No. 1
Volvo 1800 1960-1973
Volvo 120 Series 1956-1970

CONTENTS

ACKNOWLEDGEMENTS

I have noticed that in the better class of book it is customary to have a dedication. For example, my current bedtime reading is Peter Jenkins' "A Walk Across America", which is dedicated to his dog and his wife in that order. My morning relaxation is "The Old Patagonian Express" by Paul Theroux, who wishes to be remembered to his "Shanghai Lil, plus Anne, Marcel and Louis with love".

Therefore to keep up with these esteemed authors I think it is only right that this book should carry a dedication to those Morris Minor owners who have had some bearing on my life. Firstly, my Father who has owned two and who bravely loned one to me to waltz my new bride to the Continent a few decades ago. Secondly, to my daughter who was smart enough to make it her first car after she reached her majority. Next to my uncle who drives his cherished example of a well kept 1000 to this day. Fourthly to my old neighbour in Leeds, who in spite of owning much more exotic machinery still keeps a beautifully restored example in his collection. And lastly to Paul Skilleter who recently wrote the prize winning standard work for this motor classic, "Morris Minor. The World's Supreme Small Car", which has given me many hours of enjoyment.

The Brooklands series of reference books are designed to assist the owners of older interesting cars to keep them on the road by making available the articles that were written about them during their production life. Collectors, restorers, historians as well as Morris Minor owners will I am sure appreciate that the works included here are copyright and can only be reprinted in this form because of the good will and understanding of the original publishers. Our thanks in this instance go to the management of Autocar, Auto Age, Drive, Light Car, Motor, Motor Life, Motoring Life, Small Cars, Sports Car World, Thoroughbred & Classic Cars and Wheels for their ongoing support.

Available in the same series for enthusiasts is a companion volume Morris Minor 1948-1970 with a completely different selection of articles on this outstanding vehicles.

R.M. Clarke

The MORRIS MINOR Tourer

A Car which is Small in Size, Moderate in Performance but Outstanding for Economy and Charm

STAGES OF OPENING.—Fully opened to the sunshine and fresh air, the Minor displays its good turning circle, while the second photograph shows that erection of the rear sidescreens increases protection against side winds without spoiling the handsome lines of the touring body.

ASSESSED upon a basis of pleasure per £, the Morris Minor tourer must be one of the best bargains available amongst present-day cars. It is not merely an open car, in itself an extremely pleasant possession, at a price below that of any other British open tourer: it also shares with the corresponding saloon that subtle charm which can make a good small car such a delightful thing to drive.

It is now 18 months since we published a Road Test Report on the Morris Minor saloon. During the intervening period we have been able to gain continuous experience of the model's ability to cope with hard and varied usage, and at the same time increasing numbers of owners all over the world have come to understand our enthusiasm for this very unusual small car.

Inevitably, in testing the touring version of this car, comparisons are made with the saloon model upon which initial production was concentrated. The cars have individual attractions which will appeal to different groups of motorists, but it can safely be said that each is fully worthy of the other.

The saloon being a car of integral steel construction, a query naturally arises as to whether evolution of an open model, unbraced by any steel roof structure, has introduced any snags: so far as can be judged from normal driving of both models, the answer is a clear negative, roadworthiness being in no way reduced by open bodywork. There were certain rattles evident in the tourer as delivered to us for test, but the most prominent of these were so quickly silenced, by a few moments' work with a screwdriver on such items as a loose horn button and a maladjusted door catch, that the remainder, we feel confident, were caused by equally trivial details.

It is interesting to note that, despite the need for certain reinforcement of the underframe to replace the bracing effect of the steel roof, the tourer is appreciably lighter than the saloon. It will also be noted that it nevertheless shows a slightly lower absolute maximum speed, an effect which was expected in view of the impracticability of giving a folding hood contours as smooth as those of a streamlined steel roof panel.

The bulk of our experience of the Morris Minor Tourer was gained running with the hood folded away into its neat envelope and the rear sidescreens put in their bag inside the spare wheel compartment. Run thus, it is a delightful open car, not unduly troubled by the back-draught inevitable behind a large windscreen, and with wind-down glass windows and hinged ventilation panels on the doors allowing some adjustment of the amount of fresh air blowing through the car.

With the hood raised, the car becomes closely equivalent to the

LENGTH AND BREADTH.—Ample footroom for rear seat passengers is provided under the tubular frames of the adjustable driving and fixed passenger seats. There is also a very ample amount of width to allow four people to remain oomfortable during long journeys.

The Morris Minor Tourer - - - - - - - - - - - - Contd.

saloon in internal accommodation, weatherproofness and outward visibility. Entry to the back seats of a two-door body is never of the easiest, but the car is a very genuine and comfortable four-seater, giving very ample footroom and an amount of elbow width which eliminates any impression of smallness.

The sole disappointing feature of the touring body, in our view, is the fact that either raising or lowering the roof is a rather slow operation best carried out by two people. It may be tackled single-handed, and familiarity would expedite it, but we feel that in 1950 it should be possible to evolve a folding top which would be easier to operate without adding unduly to the cost.

A detail which is appreciated is the provision of locks on the doors, for use when the car is closed, and on the capacious luggage boot, which provides useful safe stowage for coats and gloves when the car is open.

Inside the body, there is a roomy glove box on the facia panel, and below the instruments a wide shelf with a retaining lip from which maps or papers do not blow away.

Exceptional Roadworthiness

On the road, the Morris Minor is outstanding as being one of the fastest slow cars in existence. Neither speed nor acceleration are exceptional, and in top gear the acceleration is frankly leisurely. There is, however, a simple four-speed gearbox which facilitates the instant use of all available engine power whenever need arises: more important, there is the sort of exceptional roadworthiness which results in maintenance of high speeds around bends and confident negotiation of small openings in traffic, the car being surprisingly seldom checked from its cruising gait.

This cruising gait is, in relation to the maximum speed of the car, unusually high, a genuine 50 m.p.h. in comparative silence as a matter of course, and 55 m.p.h. or more without protest when required. At such speeds the car rides with exceptional steadiness, the torsion bar front and leaf rear springs being extremely well damped, and holds steadily on its course while being instantly willing to alter course in response to the unusually light and sensitive steering.

Although having this delightful firmness, which makes it untiring on long day journeys, the springing system is in fact well able to absorb the shocks of potholed gravel tracks. It allows a modest amount of rolling and tyre howl to occur during fast cornering, but would be rated as excellent in all-round performance on a car of any size.

High Speeds on Third Gear

The 919 c.c. side-valve engine is extremely docile, and will happily, if not rapidly, accelerate the car from 10 m.p.h. in top gear if required. Rather unusually, it pinked on low-grade British petrol only when running at more than about 3,000 r.p.m., becoming quite prominently audible if revved hard in the indirect gears, whereas in direct drive it is little heard above the small amount of wind noise which the car makes.

Incidentally, although 30 m.p.h. is perhaps the natural speed at which to change up from third into top gear, delaying the change until 40 m.p.h. greatly improves acceleration and even a further 10 m.p.h. postponement of the change is permissible occasionally.

So pleasant to drive is the Minor, thanks to its steadiness and controllability, that a tester is tempted to forget that economy of running is its essential *raison d'être*. The fact is realized eventually, however, that although long day runs are being undertaken and the fuel tank capacity is only 5 gallons, visits to petrol filling stations are quite rare occasions. In fact, even pushing along with such vigour as Bank Holiday traffic streams across the New Forest district allowed, over 40 m.p.g. was recorded, and no abnormal restraint over cruising speed or use of the gears for acceleration is needed to produce an out-of-town fuel consumption of 50 m.p.g. As regards engine oil, 700 fast-driven miles did not lower the sump level far enough to leave room for an added pint, nor incidentally produce any indication of stiffening of the easily lubricated steering swivels.

Easy Starting

Tuned as it was for reasonable economy, the Minor's engine nevertheless started instantly after summer nights spent in the open air, and was almost immediately ready to run without use of the rich mixture control. The car carried lights adequate to its performance, although it must be admitted that on undulating roads the raised-mounting head lamps supplied to certain export markets would probably have justified their less neat appearance. Traffic indicators are fitted to the open body, but we would prefer their rather out-of-sight control to take over from the switch operating a single windscreen wiper the prominent position it occupies above the centre of the facia panel. We would also like control pedals spaced out more widely to accommodate broad country footwear, but perhaps this is impracticable with roominess obtained by extension of the body space forwards between the front wheels.

Perfection, obviously, cannot be expected of a car built to strict limits of price and size. What the Morris Minor tourer offers is practical merit in full measure, plus a very large quantity of that indefinable quality, charm.

CAPACIOUSNESS.—(Below) The large roll containing tools and wheel-changing equipment, the spare wheel, and an envelope accommodating the rear sidescreens, may be stowed in a compartment below the luggage locker. This latter is of very substantial capacity, and is provided with a lock, an appreciated detail by no means universal on open cars.

ACCESSIBILITY.—The broad bonnet, illuminated internally at night by the parking lamp bulbs, provides good access to the side-valve engine, and all auxiliaries are arranged in a spacious and correspondingly convenient manner.

The Motor Road Test No. 13/50

Make: Morris **Type:** Minor Tourer

Makers: Morris Motors Ltd., Cowley, Oxford

Dimensions and Seating

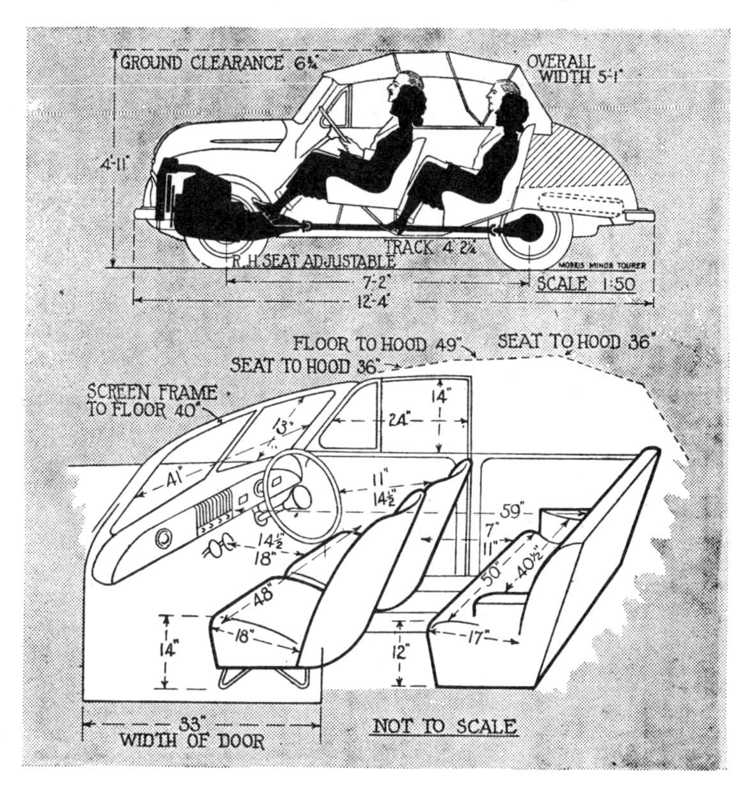

Test Conditions

Warm, dry weather with light cross wind; smooth tarmac surface; Pool petrol. Car tested with hood and sidescreens erected.

Test Data

ACCELERATION TIMES on Two Upper Ratios

	Top	3rd
10–30 m.p.h.	18.2 secs.	11.1 secs.
20–40 m.p.h.	21.3 secs.	13.1 secs.
30–50 m.p.h.	28.6 secs.	21.4 secs.

ACCELERATION TIMES Through Gears

0–30 m.p.h.	8.7 secs.
0–40 m.p.h.	15.9 secs.
0–50 m.p.h.	29.2 secs.
Standing quarter-mile	27.1 secs.

MAXIMUM SPEEDS

Flying Quarter-mile

Mean of four opposite runs	58.7 m.p.h.
Best time equals	60.0 m.p.h.

Speed in Gears

Max. speed in 3rd gear	50 m.p.h.
Max. speed in 2nd gear	36 m.p.h.

FUEL CONSUMPTION

54.0 m.p.g. at constant 20 m.p.h.
58.0 m.p.g. at constant 30 m.p.h.
50.0 m.p.g. at constant 40 m.p.h.
41.5 m.p.g. at constant 50 m.p.h.
Overall consumption for 210 miles, 5 gallons=42.0 m.p.g.

WEIGHT

Unladen kerb weight	14¼ cwt.
Front/rear weight distribution	57/43
Weight laden as tested	18 cwt.

INSTRUMENTS

Speedometer at 30 m.p.h.	7% fast
Speedometer at 60 m.p.h.	8% fast
Distance recorder	5% fast

HILL CLIMBING (at steady speeds)

Max. top-gear speed on 1 in 20	37 m.p.h.
Max. gradient on top gear	1 in 18 (Tapley 125 lb./ton)
Max. gradient on 3rd gear	1 in 10.6 (Tapley 210 lb./ton)
Max. gradient on 2nd gear	1 in 7.8 (Tapley 285 lb./ton)

BRAKES AT 30 m.p.h.

0.93 g. retardation (= 32½ ft. stopping distance) with 150 lb. pedal pressure.
0.86 g. retardation (= 35 ft. stopping distance) with 100 lb. pedal pressure.
0.54 g. retardation (= 56 ft. stopping distance) with 50 lb. pedal pressure.
0.22 g. retardation (=137 ft. stopping distance) with 25 lb pedal pressure.

In Brief

Price £299 plus purchase tax £83 16s 1d. equals £382 16s. 1d.

Capacity	918.6 c.c.
Unladen kerb weight	14¼ cwt.
Fuel consumption	42 m.p.g.
Maximum speed	58.7 m.p.h.
Maximum speed on 1 in 20 gradient	37 m.p.h.
Maximum top gear gradient	1 in 18
Acceleration:	
10-30 m.p.h. in top	18.2 secs.
0-50 m.p.h. through gears	29.2 secs.

Gearing:
14.9 m.p.h. in top at 1,000 r.p.m.
63 m.p.h. at 2,500 ft. per min. piston speed.

Specification

Engine

Cylinders	4
Bore	57 mm.
Stroke	90 mm.
Cubic capacity	918.6 c.c.
Piston area	15.8 sq. ins.
Valves	side
Compression ratio	6.6:1
Max. power	27 b.h.p.
at	4,400 r.p.m.
Piston speed at max. b.h.p.	2,600 ft. per min.
Carburetter	S.U. horizontal
Ignition	Lucas coil
Sparking plugs	14 mm. Champion L10
Fuel pump	S.U. electric

Transmission

Clutch	Borg and Beck s.d.p.
Top gear (s/m)	4.55
3rd gear (s/m)	7.015
2nd gear (s/m)	10.477
1st gear	17.994
Propeller shaft	Hardy Spicer, open
Final drive	9/41 Hypoid bevel

Chassis

Brakes	Lockheed hydraulic (2 l.s. front)
Brake drum diameter	7 ins.
Friction lining area	67.2 sq. ins.
Suspension:	
Front	Torsion bar and wishbone I.F.S.
Rear	Semi-elliptic leaf springs
Shock absorbers	Armstrong hydraulic
Tyres	Dunlop 5.00×14

Steering

Steering gear	Rack and pinion
Turning circle	31 ft.
Turns of steering wheel, lock to lock	2³/₈

Performance factors (at laden weight as tested)

Piston area, sq. ins. per ton	17.6
Brake lining area, sq. ins. per ton	75
Specific displacement, litres per ton mile	2,055

Fully described in "The Motor," October 27, 1948.

Maintenance

Fuel tank: 5 gallons. **Sump:** 6¼ pints, S.A.E. 30. **Gearbox:** 1½ pints, S.A.E. 90 gear oil. **Rear Axle:** 1½ pints S.A.E. 90 E.P. gear oil. **Steering gear:** S.A.E. 90 gear oil. **Radiator:** 13½ pints (1 drain tap). **Chassis lubrication:** By grease gun every 500 miles to 9 points. **Firing order:** 1, 3, 4, 2. **Ignition timing:** T.D.C. static. **Spark-plug gap:** 0.018—0.022 in. **Contact breaker gap:** 0.012 in. **Valve timing:** Inlet opens 8° before T.D.C.; Inlet closes 52° after B.D.C.; Exhaust opens 52° before B.D.C.; Exhaust closes 20° after T.D.C. **Tappet clearances** (hot): Inlet and exhaust 0.017 in. **Front wheel toe-in:** ³/₃₂ in. **Camber angle:** Nil. **Castor angle:** 3°. **Tyre pressures:** Front 22 lb., rear 22/24 lb. **Brake fluid:** Lockheed Orange (Overseas, Lockheed No. 5). **Battery:** 12-volt, 38 amp./hour. **Lamp bulbs:** Headlamps, N.S. 36/36 watt, O.S. 36 watt. Pilot and number plate lamps, 6 watt. Stop/tail lamp 24/6 watt. Ref. B/10/50.

WELL PROPORTIONED.—New holder of an honoured title, the Morris Minor can claim to be one of the smartest and most practical modern small cars.

1949 CARS

The MORRIS MINOR

A Handsome New Car Designed for Economy and Versatility

ECONOMY in running costs is the primary objective of the new Morris Minor, a car which is uniquely well attuned to the needs of many motorists to-day. Replacing the popular Series E Morris Eight, it offers improved fuel consumption yet also has increased body width, enhanced performance, and greater riding comfort.

The very substantial all-round progress which this newcomer signals is the result of a long period of development work, a period during which many attractive schemes were tried and abandoned before the present model was evolved. In the ultimate form, a well-tried power unit continues in use, but becomes more economical in a lighter, freer-running car with raised gear ratios. After more than a dozen years of experience with chassisless construction, Morris Motors, Ltd., have retained it for 1949, and have effected a weight saving despite the new car being roomier than its predecessor.

More important than weight reduction, however, is the shaping of the new car to pass more freely through the air. Streamlining, which, it is stated, has produced approximately 5 m.p.h. more maximum speed with unchanged power output, is even more beneficial in providing more economical running at moderate road speeds. An innovation which has assisted in the attainment of lines which are right both æsthetically and aerodynamically is the reduction of road-wheel-rim diameter to only 14 ins., tyre section being 5.00 ins.

Constructional Methods

The body, as has been said, serves also as chassis, and is of all-steel construction. An important detail point is that the box-section side members incorporate drillings which, apart from saving weight, allow both interior and exterior surfaces to receive rust-proofing treatment. Apart from the body frame lower members, two box-section longerons set at a narrower spacing extend forwards from the central cross-member of the car and pass on each side of the power unit.

The four-cylinder side-valve engine is set very far forwards in the car, so that the two-piece steering track rod actually passes above the gearbox. It is a proven power unit for which spares are universally available, and is accommodated in an exceptionally wide compartment with large opening top panel, an arrangement which provides outstanding accessibility to all parts. At night the inbuilt side lamps light up the whole interior of the engine compartment.

Independently sprung front wheels are now used, the layout being much more clever than a bare recitation of unequal transverse wishbones and hydraulically damped torsion bar springs might suggest. Mechanically simple, requiring a minimum of service attention, the design yet provides very desirable characteristics.

The lower transverse member on each side of the car comprises two channel sections set back to back to form an I-beam, the front section being a steel pressing, the rear one a forging splined to the longitudinally mounted torsion bar. Extending diagonally forward from the outer end of this composite I-beam, and acting as a tension bracing against braking stresses, is a circular-section rod, the chassis end of which is universally mounted in rubber bushes.

The upper transverse member of each wheel-locating linkage is set very high, so that the loads imposed on it are minimized, and forms the cranked actuating arm of an Armstrong hydraulic shock absorber. Steering king-pins, as such, do not exist, their duty being divided between two threaded bearings set one above the other. In every detail the I.F.S. system has been laid out not merely to perform well when in new condition, but also to wear well and require a minimum of maintenance.

The steering mechanism is simple and well conceived, comprising a rack and pinion gear behind the engine, linked to the wheels by the two halves of a divided track rod. Steering on the road is light and steady, while there is a notably compact turning circle to make the Minor a most convenient town car.

Rear suspension is by orthodox semi-elliptic leaf springs, the axle being of semi-floating construction and incorporating hypoid gears to lower the propeller shaft. An unusual point is that the Armstrong shock absorbers are mounted on the axle casing, and not on the frame of the car as is more usual, so that any noises produced will not be audible to passengers. Countless detail points

MORRIS MINOR DATA

Engine Dimensions:	
Cylinders	4
Bore	57 mm.
Stroke	90 mm.
Cubic capacity ..	918 c.c.
Piston area	15.8 sq. ins.
Valves	Side
Compression ratio ..	6.6 to 1
Engine Performance:	
Max. b.h.p. (Bare engine)	27
at	4,400 r.p.m.
Max. b.m.e.p... ..	113 lb./sq. in.
at	2,400 r.p.m.
B.h.p. per sq. in. piston area	1.71
Peak piston speed, ft. per min.	2,600
Engine Details:	
Carburetter	SU, 1-in. bore
Ignition	Coil
Plugs: make and type	14 mm. Champion L10
Fuel pump	SU electric
Fuel capacity	5 gallons
Oil filter	—
Oil capacity	5½ pints

Engine details (cont.):	
Cooling system ..	Fan and thermo-siphon
Water capacity ..	13½ pints
Electrical system ..	Lucas 12-volt
Battery capacity ..	50 amp./hrs.
Transmission:	
Clutch	6¼ ins. dry plate
Gear ratios:	
Top	4.55
3rd	7.01
2nd	10.48
1st	17.99
Rev.	17.99
Prop. shaft	Hardy Spicer
Final drive	Hypoid
Chassis Details:	
Brakes	Lockheed hydraulic
Brake drum diameter	7 ins.
Friction lining area ..	67.2 sq. ins.
Suspension:	
Front..	Independent by torsion bars
Rear	Semi-elliptic
Shock absorbers ..	Hydraulic double-acting
Wheel type	Pressed type
Tyre size	5.00 × 14, Dunlop

Chassis Details (cont.):	
Steering gear	Rack and pinion
Steering wheel ..	16½ ins. spring
Dimensions:	
Wheelbase	7 ft. 2 ins.
Track:	
Front	4 ft. $2\frac{5}{16}$ ins.
Rear	4 ft. $2\frac{5}{16}$ ins.
Overall length ..	12 ft. 3½ ins.
Overall width.. ..	5 ft.
Overall height ..	4 ft. 10 ins.
Ground clearance ..	6¾ ins.
Turning circle.. ..	35 ft.
Dry weight	14¾ cwt.
Performance Data:	
Piston area, sq. ins. per ton	21.4
Brake lining area, sq. ins. per ton	91
Top gear m.p.h. per 1,000 r.p.m... ..	14.9
Top gear m.p.h. at 2,500 ft./min. piston speed..	63
Litres per ton-mile, dry	2,500

like this have received attention in design stages, in the effort to evolve a popular-priced car which will remain quiet long after the first flush of youth has passed.

The new 2-door saloon body is a comfortable four-seater of very adequate length and height but especially notable for interior width. Individual front seats are provided, and the low level of the sunken floor allows very adequate toe space for rear-seat passengers. Although a simple gear lever of orthodox design is retained, the power unit is set so far forward that this lever does not obstruct entry to the driving seat from either door.

The tail of the car incorporates a luggage locker with external lift-up lid, the spare wheel and tyre being accommodated horizontally in a lower tray. A rare feature is a large opening between car and luggage locker, so that on occasion the rear seat squab may be folded forwards to throw together locker and rear compartment and form a single luggage space of large capacity.

In addition to the usual wind-down panels, the front windows incorporate hinged sections at the forward edges to provide ventilation without draughts. The windscreen is of V form, but the attainment of low wind resistance has not involved sacrifice of vision through the rear window. Forward vision, over the short, rounded bonnet, is exceptionally good.

We have been privileged to drive one of these cars on the road, and were impressed by its smooth behaviour. It rides comfortably over varied surfaces, yet shows pleasantly little roll on corners, and has light and steady steering to which it responds willingly. Dazzling performance is not compatible with extreme economy, but there is brisk acceleration if the four-speed gearbox is used freely, and we have no reason to doubt the Minor's ability to exceed the mile-a-minute mark on open roads.

A smart open tourer model will also appear on the Morris stand at Earls Court.

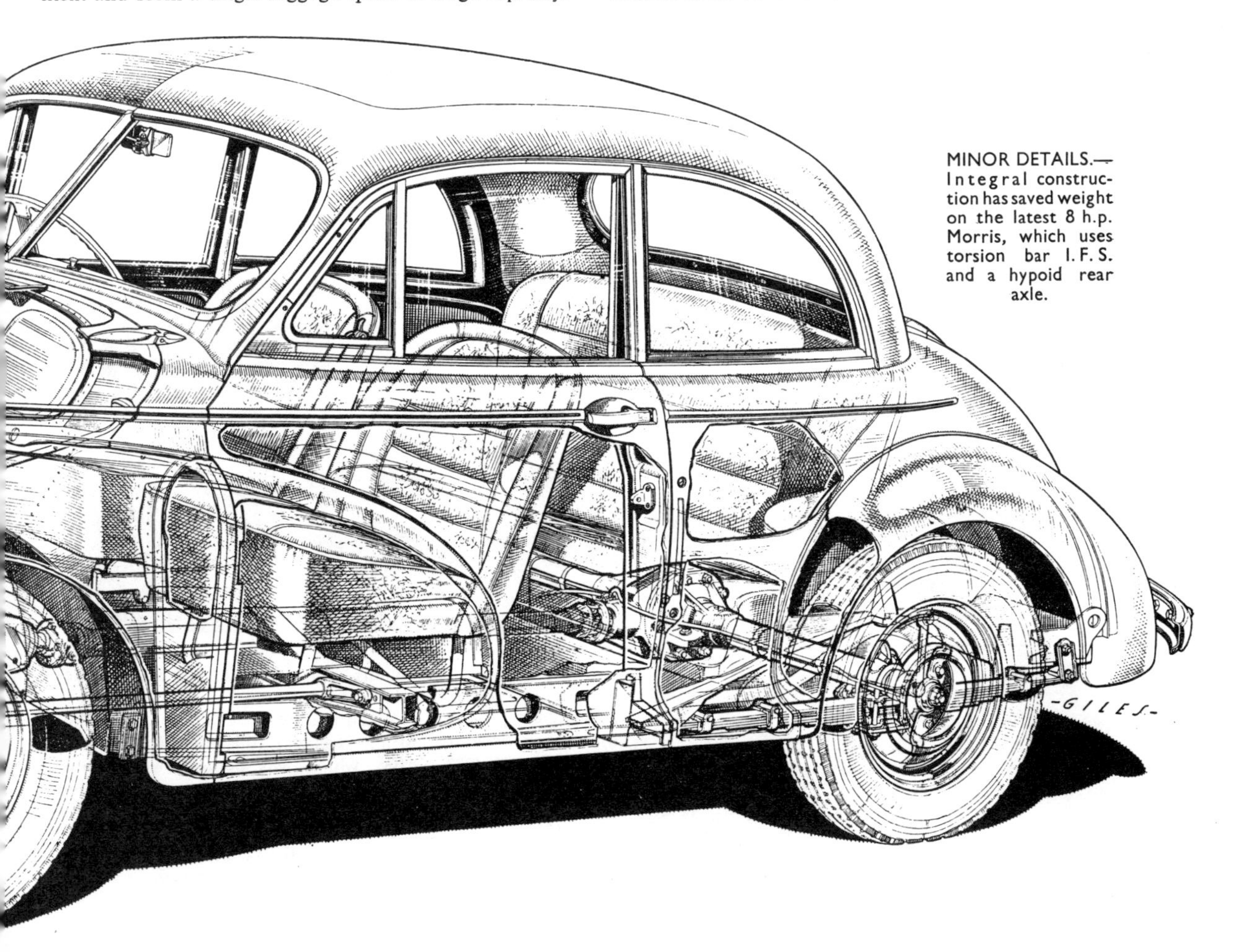

MINOR DETAILS.—Integral construction has saved weight on the latest 8 h.p. Morris, which uses torsion bar I. F. S. and a hypoid rear axle.

RUNNING ANTI-CLOCKWISE around Goodwood track, the o.h.v. four-door Minor is seen here passing the stands.

10,000 MILES in 10 DAYS by MORRIS MINOR

Ambitious Nuffield Non-stop Test-run Using Ingenious Mobile Servicing Method

NOT seeking to set up International records, but to carry out an extremely searching road test under constant scientific supervision, the Experimental Department of the Nuffield Organization last week did what has never been done before.

A series production Morris Minor four-door saloon, with the latest o.h.v. 803 c.c. BMC engine (first fruits of the arrangement between the Austin and Nuffield organizations) was driven for 10 days completely non-stop, covering 1,000 miles per day.

The run commenced, not on a circular banked track where speeds can be kept constant, but on the well-known Goodwood road circuit, where corner succeeds corner, on October 1 at 6 a.m. and finished at about 11 a.m. on October 10. The little car, which is an "export only" model, was driven by a team of six drivers, taking turns at the wheel, and was averaging all day and all night something over 45 m.p.h., at a fuel consumption which grew steadily better after the first 4,000 miles and at no time fell below 40 m.p.g.

Naturally, the car was well prepared for its task of covering a year's motoring in 10 days, but except for a larger petrol tank and special seat hinges to allow drivers to clamber in and out of the car while it was still in motion, nothing was altered on the vehicle.

Where this performance differs from normal record attempts (which this was not) is that throughout the run the engine never ceased driving the car and the wheels never stopped rotating.

This remarkable feat was achieved by the use of a truck and an articulated trailer as a tender. The trailer was in the form of a hollow cage on wheels, with platforms at the sides where mechanics could work and a frame-superstructure overhead, floodlit at night, and coupled to the pilot vehicle so that the Minor's "pulse and temperature" could be taken by technicians sitting at instruments in the pilot car.

When maintenance was required, the Minor was carefully driven into the cage, which was being towed round the circuit. When once within, bars were dropped into place front and rear, and a series of lights told the driver that he was still keeping his car under power, and not being towed.

TAPE RECORDERS were used to register two-way V.H.F. radio conversations between technicians in the mobile tender and the static control room.

TIMING AND CONTROL were undertaken in what is usually the race timekeepers' box. Laps completed were automatically recorded each time the car crossed a strip across half the track.

THREE STAGES in mobile servicing technique: Above, the Minor enters a marked "lane" and drives up to the rear of the moving tender. When the car is in the bay with the rear struts in position (left) light signals tell the driver when the car is exerting positive thrust on the tender. Below, oil, water and tyres are attended to inside the servicing bay without the Minor ceasing to propel itself. Rear-wheel changes were done by lifting clear and braking the "change" wheel and driving with the other.

Located thus, and with the hand throttle set, the car was left to the mechanics. Fuel and oil were put in, and by raising one wheel at a time on a block and tackle (there being hand brakes to each rear wheel) the wheels could be changed while the car was still motoring itself at some 15-20 m.p.h. and the non-stop run was uninterrupted.

A very detailed record was kept of the car's performance, its fuel consumption and its tyre wear, which, on the Goodwood circuit with its corners, were far higher than would occur during a 10,000-mile run on normal roads. Indeed, the engineers discovered that appreciable horse-power was consumed in the constant cornering alone, and petrol consumption was elevated by nearly 8 m.p.g. in the process—equivalent to covering 15 per cent. more mileage on a straight road and requiring 807,800,000 ft./lb. greater energy output! Tyre wear was in ratio. Tyres were worn with the constant cornering, day and night, at six times the normal rate of road wear.

In actual figures, it was found that at the set average speed of 45 m.p.h., tyre wear on the front wheels was at the rate of 1 mm. of tread depth per 250 miles on the Goodwood circuit. This compares with 1 mm. per 1,500 miles at the same speed on ordinary A and B class roads.

In normal record-breaking runs of this nature, of course, the car stops at the pits while maintenance is carried out, but in this achievement, the Morris was never stopped at all, and the reliability and endurance of the engine were stressed to the maximum possible.

The run was extended to slightly over the exact 10 days, providing a margin to allow for possible errors in measurement. In all, 10,148 miles (4,264 laps) were covered at an average speed of 45.3 m.p.h. and with an average fuel consumption of 43 m.p.g.

10,000 Miles Wi[…] The M[…]

Expe[…]
with[…]

THE "£100" Morris Minor with side-valve engine is being retained practically unchanged for 1932, and an account of 10,000 miles' running with a staff-owned 1931 model of this type will reveal its many outstanding and attractive features.

The model in question—illustrated on these pages—is the "£100" chassis fitted with the standard open tourer body—costing £112 10s.—and has been given some strenuous work to do since April of this year.

First, a few brief particulars of the chassis. The side-valve engine has a bore and stroke of 57 mm. and 83 mm.—847 c.c.; tax £8. The wheelbase is 6 ft. 6 ins., the track 3 ft. 6 ins., and there are 8½ ins. ground clearance. The carburetter is an S.U. and coil and battery ignition is employed.

Transmission is by a single-plate clutch, three-speed gearbox and open propeller shaft with two fabric universal joints. The springs both fore and aft are semi-elliptic.

In choosing this particular model the owner was guided by two desiderata—the car must be of the open variety and there must be room for occasional third and fourth passengers. The open four-seater is ideal for these purposes, and when travelling alone, or two up, the room in the rear compartment is sufficient for plenty of baggage.

In Pursuit of Luxury.

On taking delivery of the model the six-foot-tall owner decided that, as very long journeys by day and night were to be frequent, maximum comfort must be sought, and although the standard bucket seats of the car appeared perfectly comfortable, a predisposition towards pneumatic upholstery led to their removal.

Two pneumatic "buckets," obtained from the Abbey Coachworks, High Path, Merton, S.W.—were substituted. They were simply dropped into the Morris on the standard fixing—where they fitted without alteration—and, of course, they hinge forward to give access to the rear seats.

The next modification—still in pursuit of £1,000 comfort for £112 10s.!—was to lengthen the steering column by 8 ins. This was done simply by removing the steering wheel, screwing on an extension obtained from the Abbey concern and replacing the wheel—this time a large spring-spoked "Dover" sports.

Next, the rake of the steering column was lowered,

ANYWHERE IN ANY WEATHER. — (Top, left) The Morris Minor w[…] gradient—a tribute to the brak[…] covered in five months. It is a […] warm and comfort[…]

necessitating turning the track-rod upside down and making up a long clip for the facia board. An ideal driving position for a very tall driver resulted.

The car was carefully run in, and the usual changing of the oil in sump, gearbox and back axle was carried out at 500 miles.

After the running-in period the Morris was called upon to work hard and for long periods with but scant attention, and this the little car has willingly done.

The power unit is noticeably silent and sweet; there

RIS MINOR

over a Long and Strenuous Period r-seater Edition of the £100 Morris. oothness and Reliability Outstanding tures of an Attractive Little Car

equipment in place and (centre) safely parked on a 1 in 4 The car has given trouble-free service for 10,000 miles, achine, reliable, with a pleasing turn of speed, and very od and bad weather.

is no tappet clatter and no vibration. When cruising at a steady 40-45 m.p.h.—a normal gait—the engine is perfectly happy, runs cool and never tires.

On the level the maximum speed is about 55 m.p.h., with 40 m.p.h. on second and 20 m.p.h. on first. The steering, which is of the low-geared type, has always been very light, and there is no fatigue in driving the car over long distances. Road holding is good, there is no tendency to leap and "bucket" from side to side at speed, and there is no suggestion of kick in the steering wheel even when traversing bad surfaces.

The brakes on the 1931 Morris are a great improvement on previous types, and although very heavy use has been made of them during the past 10,000 miles, there is still no need for relining. Only three times in this not inconsiderable distance has adjustment been necessary.

Nothing phenomenal has been accomplished in the way of average speeds, but with constant regularity long journeys to the North and West of England have been made in a running time not to be despised by the owner of a large car; and on several occasions journeys of over 200 miles have been made running in company with large cars, in which the latter have travelled at their normal touring gait—and the little Morris, without fuss or strain, has kept its station.

170 Miles at 35 m.p.h.

From Exeter to London, for example, an average speed—exclusive of timed stops—of 32 m.p.h. and 35 m.p.h. have been accomplished quite easily, and higher speeds have been registered on the Great North Road, where 15 miles were once covered at an average of 42 m.p.h.

Probably the outstanding feature of the car is its unfailing reliability. Every day it covers a minimum of 50 miles—to and from London. Every morning and evening it is called upon to start up immediately and without further ado carry its owner to the office or home again.

During all this time the starting handle has never been used. Day in, day out, for week after week, this has been the invariable programme, and at week-ends, without any preparation other than a ten-minute "grease up," the Morris has been driven hard half across England and back in the course of duty.

Never once in this arduous existence has the car let its owner down—and apart from one decarbonization (at 9,000 miles, be it noted!)—a spanner has never been put on the chassis.

The plugs have been disturbed only once—when the head was off—the piston rings have not seen daylight yet, and the tappets went untouched until the recent "decoke" was undertaken; in fact, if ever a car has functioned with a maximum of efficiency on a minimum of "maintenance," it is the Morris illustrated on these pages!

Greasing has been carried out regularly, and in addi-

(Below) Showing the surprising roominess of the rear seats where deep foot wells are provided. Note the special pneumatic cushions which were fitted for maximum comfort.

tion to the weekly (i.e., every 400 miles) greasing, once a fortnight the clutch race has received lubricant, the floorboards have been lifted and the brake gear oiled.

There is a point of criticism which may be mentioned in this connection, and this is the inaccessibility of the grease nipple which serves the shaft upon which both the clutch and brake pedal rock.

With the gun supplied as standard it is impossible to reach the nipple from under the bonnet, and raising the driver's toe-board proves a long and laborious process, necessitating the disconnection of throttle and ignition controls and some 20 minutes' work. A right-angle type of nipple, pointing towards the front, might be a solution.

No trouble has been experienced with carburation or ignition, and the dynamo continues to turn out a lusty 10 amps. on full charge. In this connection it may be mentioned that the headlamps provide an adequate driving light.

No criticism can be lodged against the all-weather equipment. The hood is a sturdy piece of work, easily put up or down single handed, and the side curtains fit quite as well as side curtains ever do. A point which might be remedied is the scant space provided behind the rear-seat squab for stowing the curtains when out of use. The hood furls neatly and is encased in a black envelope, and as a personal whim—to keep dust from the rear seats when driving solo or two up—a tonneau cover has been added.

The Dunlop tyres have given faithful service during the period under review. The photographs illustrating this article were taken after a further 3,000 miles had been covered, and one of the original tyres may be seen on the spare wheel—not by any means worn out.

Here, then, after 10,000 very hard miles, is a sturdy little car going as sweetly and gamely as ever, with real comfort for two and room for four, a petrol consumption of round about 40 m.p.g. under hard driving and an oil consumption of 1,500 m.p.g. What more could anyone want for a modest £112 10s.?

MORRIS MINOR

(*Continued from Page 41*)

through gears from 0 to an indicated 50 mph the time was consistently less than 20 seconds—19, 19.5 and 18.5 seconds to be exact. This to me was very good on the level ground and I took the Minor to a grade about one-quarter-mile in length that I had been told was 22%. I stopped at the foot of the hill and, going through gears, this mighty little midget took me to the crest of the hill, after changing into top gear just before reaching it, in about 34 seconds. These facts astounded me.

I was not looking for sports-car performance, as my requirements for a car is one that can cruise fast enough with the family aboard and have enough go to get out of the way of the big ones on the turnpike. Also, I feel a car should get me places at normal speeds and in normal times without having to rev up or continually downshift to make hills. I did not believe that a 1000 cc engine could do this with four people in the car, but the Minor proved otherwise.

Insofar as sports-car performance goes, to me it's a sort of taut responsiveness in steering and gear-changing, along with a certain glued-to-the-road feeling on fast corners. Then, if a skid should occur, it naturally should be the rear. Throttle response and steering should then be such as to allow quick corrections. The Morris has these qualities because on alternately wet and dry pavements, besides fast cornering, I was able to sort of "set up" skids and quick recoveries.

Apart from all its other attributes, let me say here that the Morris Minor 1000 comes with built-in insurance in its steering and stability. However, like any other car, it is not without its imperfections—but in this particular instance they are so insignificant as not to be held against this splendid little vehicle.

On fast going over bad surfaces there was an occasional trace of kick-back in the steering. But of course, you can't have your cake and eat it too. I also feel that the instruments on the dashboard should be placed directly in front of the driver—not in the center of the dash.

Adjustment of the front seats could be improved by increasing the movement a little further forward. Even when fully forward it is necessary for a person of smaller stature to lean out front to reach the gear shift. This change would improve the leg room for passengers in the back seat which, in my opinion, is not sufficient for comfort on really long hauls.

These were about the only disadvantages that I could find in this otherwise superb little automobile.

Interior decor is not of the lady's boudoir type, but simple and practical, with a very good washable synthetic leather throughout.

As far as could be estimated, this car should give about 45 mpg at a steady 30 mph, but this consumption should drop to about 30 to 35 mpg at road speeds of 60 mph with four passengers.

I believe my next new car will be a Morris 1000. It has everything I want: comfortable driving for long stretches, excellent accelerator response, good stability without loss of riding comfort, economy, and fast travel when necessary.

I don't know what the top speed of this car is, but believe it to be in excess of 70 mph. This top speed capability should assure prolonged life and mileage when operating the engine at normal cruising speeds. Furthermore, this reserve of power and speed, easily attained, indicates safe passing when necessary.

How one can expect so much from a small wagon costing $1,695? The only answer I can give is: "seeing is believing." I've stated the straight facts.

CONTINUED FROM PAGE 44

checked for accuracy at 30 m.p.h. only, and at this indicated speed we found the car was doing 31.3 m.p.h.

Outstandingly Flexible . . .

Flexibility was outstanding. We could drop as low as 14/15 m.p.h. in high gear, and then, accelerating, draw smoothly away. In town use, high gear proved adequate for most traffic "loafing," whilst a quick chop to third laid on acceleration strong enough to cope with most 10-30 m.p.h. eventualities—or, in dire emergency, second gear used intelligently was positively sporty!

It was also delightful to be able to rush up to a corner headlong, go for the brake, and with a quick rock-of-the-foot be able to hoist the motor up sufficiently high for a slip-through into third without having to ease the pressure one iota on the brake pedal! Most people these days would *like* to toe 'n' heel—they even *try* to toe 'n' heel—*but few are the cars nowadays, with all our modernity, which will gracefully permit of this handy and very elementary manoeuvre.*

We enjoyed our sojourn in the little V. & L. stationwagon. It is a pleasant, well-mannered little car, is solidly put together, and—we feel—at the price, represents honest value as station wagons go today.

We were sorry to have to return it, and look forward with some eagerness to trying out the first local stationwagon conversion of the "1000," which is due out soon. Doubtless it will have some points that are more desirable than some of those of the 8 h.p.; but we feel that there are others, such as the truck-type springing, solid chassis, and low, practical gearing that it will not have.

But it is certain that there is a ready place for both vehicles in the scheme of things today. ●

The Autocar ROAD TESTS

The Morris Minor has a well-balanced appearance, and the four doors provide easy access to the comfortable and light interior. The head lamps are now larger in diameter and mounted in the wings.

DATA FOR THE DRIVER

MORRIS MINOR

PRICE, with four-door saloon body, £365, plus £204 5s 7d British purchase tax. Total (in Great Britain), £569 5s 7d.

ENGINE : 8 h.p. (R.A.C. rating), 4 cylinders, side valves, 57 × 90 mm, 918.6 c.c. Brake Horse-power : 27.5 at 4,400 r.p.m. Compression Ratio : 6.6. to 1. Max. Torque : 38.75 lb ft at 2,400 r.p.m. 15 m.p.h. per 1,000 r.p.m. on top gear.

WEIGHT (running trim with 5 gallons fuel) : 16 cwt 0 qr 0 lb (1,792 lb). Front wheels 55 per cent ; rear wheels 45 per cent. LB per C.C. : 1.95. B.H.P. per TON : 34.4.

TYRE SIZE : 5.00—14in on bolt-on steel disc wheels.

TANK CAPACITY: 5 English gallons. Approximate fuel consumption range, 35-40 m.p.g. (8.1-7.1 litres per 100 km).

TURNING CIRCLE : 35ft 0in (L and R). Steering wheel movement from lock to lock : 2⅔ turns. LIGHTING SET : 12 volt.

MAIN DIMENSIONS : Wheelbase, 7ft 2in. Track, 4ft 2⅝in (front) ; 4ft 2 5/16in (rear). Overall length, 12ft 4in ; width, 5ft 1in ; height, 5ft 0in. Minimum Ground Clearance : 6¾in.

ACCELERATION

Overall gear ratios	From steady m.p.h. of 10-30 sec	20-40 sec	30-50 sec
4.550 to 1	23.5	23.4	36.8
7.015 to 1	13.1	14.2	—
10.477 to 1	8.7	—	—
17.994 to 1	—	—	—

From rest through gears to :—

	sec
30 m.p.h.	9.8
50 m.p.h.	38.5

SPEEDS ON GEARS

(by Electric Speedometer)	M.p.h. (normal and max)	K.p.h. (normal and max)
1st	12—19	19—31
2nd	22—31	35—50
3rd	38—46	61—74
Top	61	98

Speedometer correction by Electric Speedometer :—

Car Speedometer		Electric Speedometer m.p.h.
10	=	10.5
20	=	19.0
30	=	27.0
40	=	36.0
50	=	45.0
60	=	54.0
67	=	61.0

WEATHER ; Dry, cold ; fresh wind.

Acceleration figures are the means of several runs in opposite directions.

Described in " The Autocar " of October 13 and 20, 1950.

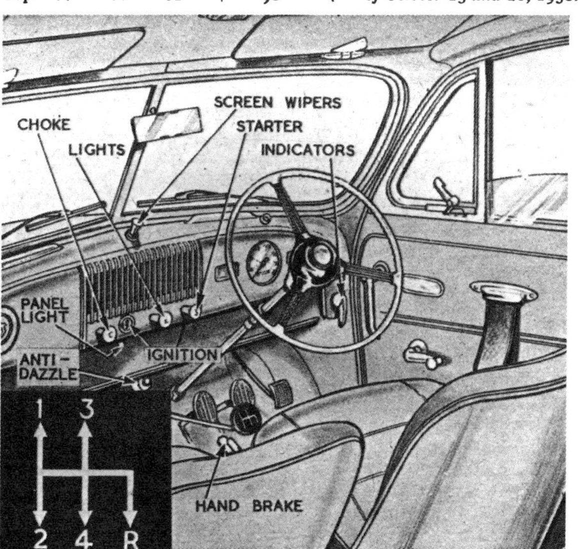

No. 1429 : MORRIS MINOR FOUR-DOOR SALOON

THE Morris Minor was a "winner" from the start, and has established itself as a most successful small car of outstanding qualities. Perhaps to a greater extent than has ever been seen previously, it is really a scaled down big car, and not only has it the great appeal of economy of running costs resulting from its modest engine size, but also the riding and control, arising a good deal from its design of torsion bar independent front suspension, cause experienced drivers to enthuse over it, even though versed in bigger and faster cars.

It has dual appeal. It is pre-eminently the small family car capable of conveying four people at a high standard of motoring comfort, and also, as a separate facet of its character, it can be driven flat out within its limits by the more enterprising style of driver, who enjoys himself because of its good handling and light but accurate steering.

Previously the two-door saloon, with which the current Series MM began, and the open tourer have been the subject of Road Tests, and now comes the latest version, the four-door saloon, which was announced just before the last London Show. Unfortunately it is necessary to prefix further comment on the four-door model with the remark that for the present it is for export only. The two-door is a thoroughly worthy little motor car and, incidentally, is now coming through with a frontal appearance similar to that which the four-door Minor has had from the start, the head lamps being faired into the wings instead of being recessed into the grille, separate side lights also being fitted. Another important new feature is the fitting of a water pump, which permits an interior heater to be made available as optional equipment. The four-door model incorporates a number of interior bodywork details that are worth having.

The four-cylinder side-valve engine is remarkably willing, in the main very smooth indeed, apart from one or two lesser tremors at various points in the range, and is able to keep the car moving at a comfortable 45 to 50 m.p.h. sustained rate, considerably higher readings being possible without appearing to cause overstress. The fact that top gear ratio is fairly high for a small engine results in mechanical fussiness being minimized, yet top gear flexibility is good and the gears do not have to be used a great deal in country of average character unless the driver is bent on obtaining the utmost from the car. In point of fact there is an excellent gear change operated by a central "prong" lever which is refreshingly direct and positive, and the engine can be revved on third and second in a smooth purr that is pleasing rather than otherwise. Few cars have ever

The small horizontal grille and the new, higher mounted head lamps together produce a pleasing frontal appearance. Separate side lamps go with this styling modification. These frontal changes were first seen in the "export only" four-door Minor, and now apply to the two-door model as well.

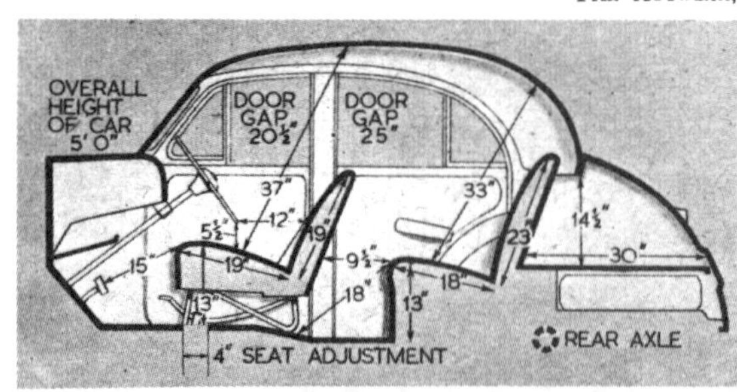

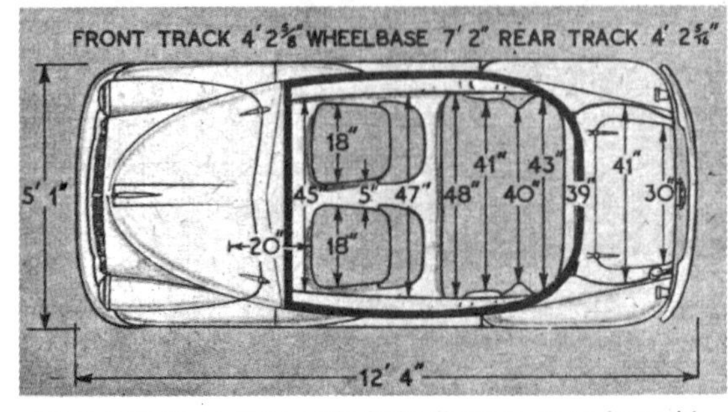

Measurements in these scale body diagrams are taken with the driving seat in the central position of fore and aft adjustment and with the seat cushions uncompressed.

felt so entirely "right" and satisfactory, in relation to size, as the Morris Minor does.

A good average speed can be made without forcing the car unduly. If it is driven fairly hard it will be a surprisingly short distance behind a much faster car on a run over English roads. On gradients speed begins to fall off fairly early on top gear, and as regards quite severe hills, in the 1 in 6-7 category (approximately 16 per cent gradient), a drop to first gear is required. There is the feeling, however, of useful power being available and of ability to tackle practically anything in the way of gradients provided that the driver knows what he is doing with a gear box. No car could be handier in difficult country with many bends and narrow roads, and the convenience resulting from modest overall dimensions for manœuvring and parking in dense traffic is appreciated; the steering lock is excellent. The more it is driven, the more one realizes what a useful and likeable little car it is.

So good is the rack and pinion steering, finger light yet

ROAD TESTS . . . . continued

completely devoid of vagueness or need to fiddle with it, that one pays it the compliment of feeling that it would well suit a car of much higher performance. It has marked castor action, no road wheel shocks come back to the steering wheel, when the tyre pressures are set to recommended figures, and in every way it is what steering should be. Largely because of its excellence one can feel completely at home in the car after only a few minutes and able to steer it with accuracy much above the average.

The suspension is firm and the car feels taut, yet the riding is good over a variety of surfaces. The Minor can be taken round bends quite fast with marked absence of roll, a feature which derives, no doubt, from the weight distribution and location of the wheels at the "corners of the car." Also, tyre squeal is absent. The four-door saloon has been tried on this occasion with a wide variety of loading, from the driver alone to a full complement of four heavy occupants, and in all circumstances it has given great satisfaction, evoking enthusiasm from the experienced and the lay mind alike.

A full load does not pull the performance down as noticeably as might be expected and even under such conditions it needs only a mildly favourable gradient for readings towards the 70 mark to be seen on the speedometer, which becomes somewhat optimistic at the higher speeds.

In addition to the usual testing on which these comments are based the car was given an exceptionally severe long-distance run in which more than 1,000 miles were compressed into a weekend trip, much of it over snow-covered roads in the North of England and Scotland.

The driving position well suits a variety of drivers, and good support is given by the slight curvature of the back rest of the separate seat, which is adjustable, though the front passenger seat is not. The steering wheel is of quite large diameter, spring spoked, and at a comfortable angle. The pull-up hand-brake lever is placed ideally between the front seats and is powerful for holding the car on a gradient. The clutch pedal action is particularly light, a remark that applies, indeed, to all the controls.

The Lockheed hydraulically operated brakes are entirely

The traditional "prong" gear lever is retained on the Minor, which has a simple yet business-like interior. A useful tray is provided below the facia, and a glove locker with press button operated lid is arranged on the passenger side. The heater, seen centrally under the facia, is optional equipment.

Easy access to the rear seats is provided by the forward opening doors. Arm rests are fitted to the rear doors and an ash tray is fitted into the propeller-shaft tunnel.

By splitting the rearward side windows in the vertical plane a much larger opening is allowed for the front portion than would otherwise be possible. The front windows are fitted with swivelling quarter lights. Both the bonnet and the luggage lid have external hinges ; the door handles are of pull-out type, with depressions in the door panels for finger clearance.

The luggage locker is of quite considerable proportions for a small car, and is clear of obstructions such as fuel filler pipes. A separate lower compartment houses the spare wheels and tools.

satisfactory, light pedal pressure producing the required results right up to the emergency stage. A slight shudder is experienced sometimes in starting from rest, and occasionally after a gear change, unless some care is given to the clutch engagement. Ordinarily quick changing, upwards and downwards, can be made without over-riding the synchromesh on second, third and top gears of the four-speed box. Some drivers find the clutch and brake pedals closer together than they like, and the throttle pedal is not entirely comfortable, especially to a long-legged driver, on a long journey.

Outward vision is good. The left wing cannot be seen by an average-height driver in a right-hand drive example, though this is not important with a car of the compact dimensions of the Minor. A tall driver may find the top rail of the V windscreen tending to cut his line of vision, and it would be better if the main windscreen pillars were thinner. There is, however, a good feeling of light within the car, and also of width. A useful view is given by the driving mirror. The horn note is not worthy of so very good a small car, and is inadequate on many occasions. In conjunction with a 12-volt battery, the new type larger head lamps give a beam fully adequate to the speed, and in fact this is as good as is found today on a number of cars of considerably higher potential performance.

This Minor four-door saloon is admirably equipped in detail and it is difficult to point to a single item omitted that could reasonably be expected in a car of this price. There are wide sun vizors, a lidded cubby hole of useful size in the facia and a full-width shelf beneath the facia, pivoting ventilator windows in the forward doors, and ash trays recessed into the two front doors and another one set into the propeller-shaft tunnel, in a position useful to the back passengers. Pulls are fitted on the front doors for closing them from inside. Three doors can be slam locked and the driver's door locks with the ignition key. There are elbow rests on the rear doors. The upholstery is of good quality. There is really useful luggage accommodation, and if rear passengers are not carried there is the practical point that the back of the seat can be folded down, thus considerably increasing the luggage space. A roof light is mounted above the windscreen, in what proves to be a very convenient position for map reading and general interior illumination. With the side lamps switched on an unobtrusive green light illuminates the ignition keyhole.

Twin windscreen wiper blades are fitted. A clock is not provided as standard, but there is provision for easy installation of one as an extra. The traffic signals are not of the usual self-cancelling type but incorporate a time-basis return mechanism in a switch below the facia, in front of the driver. An ammeter is not fitted, but there is an oil pressure gauge.

Most luxurious of all among the equipment of a small car is the new Smiths heater, available as a "standard extra"; ducts for windscreen demisting are built in. Although it is not of the type that takes in fresh air from outside, it is a real acquisition in a car of this type and proves remarkably efficient, almost too efficient at the maximum position provided by the control switch, which is a fault on the right side, as the degree of heating can be regulated by means of the switch controlling the heater fan.

From cold the engine fired at once, though sometimes in colder weather it did not keep running without a further application of the starter. The mixture control has several positions at which it can be locked, from full rich back to normal, and it is possible to let the engine warm up at a fast tick-over in a way that is not usual with modern cars.

Lifting the bonnet adequately exposes the engine for all normal requirements. The cable operated starter switch is mounted to the left of the 12-volt battery, and all ignition components are carried high up. The wide, low radiator is fitted with a pressurized cap to prevent loss of coolant.

Road Test No. 1/53—Th

Make: Morris **Type:** Minor (Series II) 4-door Saloon
Makers: Morris Motors, Ltd., Cowley, Oxford.

Dimensions and Seating

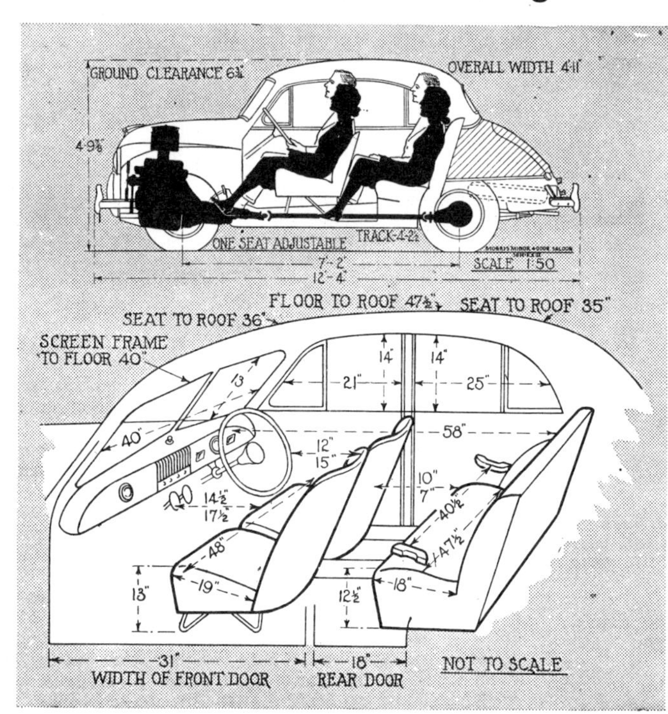

Test Conditions

Cold, damp weather with little wind. Damp tarmac surface with occasional lumpy ice. Pool petrol.

Test Data

ACCELERATION TIMES on Two Upper Ratios

	Top	3rd
10-30 m.p.h.	15.9 secs.	8.7 secs.
20-40 m.p.h.	17.6 secs.	11.5 secs.
30-50 m.p.h.	24.1 secs.	—

ACCELERATION TIMES Through Gears

0-30 m.p.h.	8.4 secs.
0-40 m.p.h.	14.9 secs.
0-50 m.p.h.	28.6 secs.
Standing Quarter Mile	26.5 secs.

MAXIMUM SPEEDS

Flying Quarter Mile

Mean of four opposite runs	62.3 m.p.h.
Best time equals	64.7 m.p.h.

Speed in gears

Max. speed in 3rd gear	44 m.p.h.
Max. speed in 2nd gear	30 m.p.h.
Max. speed in 1st gear	19 m.p.h.

FUEL CONSUMPTION

52.5 m.p.g. at constant 20 m.p.h.
52.5 m.p.g. at constant 30 m.p.h.
48.5 m.p.g. at constant 40 m.p.h.
41.5 m.p.g. at constant 50 m.p.h.
35.0 m.p.g. at constant 60 m.p.h.

Overall consumption for 472 miles, 12 gallons, equals 39.3 m.p.g.

WEIGHT

Unladen kerb weight	15¾ cwt.
Front/rear weight distribution	57/43
Weight laden as tested	19¼ cwt.

INSTRUMENTS

Speedometer at 30 m.p.h.	5% fast
Speedometer at 60 m.p.h.	7% fast
Distance recorder	2% fast

HILL CLIMBING (at steady speeds)

Max. top gear speed on 1 in 20	44 m.p.h.
Max. top gear speed on 1 in 15	30 m.p.h.
Max. gradient on top gear	1 in 14.9 (Tapley 150 lb/ton)
Max. gradient on 3rd gear	1 in 8.6 (Tapley 260 lb/ton)
Max. gradient on 2nd gear	1 in 6.3 (Tapley 350 lb/ton)

BRAKES at 30 m.p.h. (dry tarmac road)

0.97g retardation (=31 ft. stopping distance) with 125 lb. pedal pressure.
0.93g retardation (=32½ ft. stopping distance) with 100 lb. pedal pressure.
0.73g retardation (=41 ft. stopping distance) with 75 lb. pedal pressure.
0.51g retardation (=59 ft. stopping distance) with 50 lb. pedal pressure.
0.15g retardation (=200 ft. stopping distance) with 25 lb. pedal pressure.

In Brief

Price £405, plus purchase tax £226 10s., equals £631 10s.

Capacity	803 c.c.
Unladen kerb weight	15¾ cwt.
Fuel consumption	39.3 m.p.g.
Maximum speed	62.3 m.p.h.
Maximum speed on 1 in 20 gradient	44 m.p.h.
Maximum top gear gradient	1 in 14.9
Acceleration: 10-30 m.p.h. in top	15.9 secs.
0-50 m.p.h. through gears	28.6 secs.

Gearing:
13.1 m.p.h. in top at 1,000 r.p.m.
65.5 m.p.h. at 2,500 ft. per min. piston speed.

Specification

Engine

Cylinders	4
Bore	58 mm.
Stroke	76 mm.
Cubic capacity	803 c.c.
Piston area	16.4 sq. in.
Valves	Pushrod o.h.v.
Compression ratio	7.2/1
Max. power	30 b.h.p.
at	4,800 r.p.m.
Piston speed at max. b.h.p.	2,400 ft. per min.
Carburetter	Inclined S.U.
Ignition	Lucas coil
Sparking plugs	Champion NA8
Fuel pump	S.U. electrical
Oil filter	AC or Purolator

Transmission

Clutch	Borg & Beck s.d.p.
Top gear (s/m)	5.286
3rd gear (s/m)	8.88
2nd gear (s/m)	13.69
1st gear	21.62
Reverse	27.38
Propeller shaft	Hardy Spicer open (needle roller Universals)
Final drive	Hypoid bevel

Chassis

Brakes	Lockheed hydraulic
Brake drum diameter	7 ins.
Friction lining area	67.2 sq. in.

Suspension:
Front: Torsion bar and wishbone I.F.S.
Rear: Semi-elliptic leaf.
Shock absorbers:
Front and rear: Armstrong hydraulic
Tyres ... Dunlop 5.00x14

Steering

Steering gear	Rack and pinion
Turning circle	33 ft.
Turns of steering wheel, lock to lock	2¼ ft

Performance factors (at laden weight as tested)

Piston area, sq. in. per ton	17.0
Brake lining area, sq. in. per ton	70
Specific displacement, litres per ton mile	1,910

Fully described in "The Motor," October 15, 1952

Maintenance

Fuel tank: 5 gallons. **Sump:** 5 pints, S.A.E. 30 summer, S.A.E. 20 winter. **Gearbox:** 2¼ pints, S.A.E. 40. **Rear axle:** 1½ pint, S.A.E. 90 Hypoid gear oil. **Steering gear:** S.A.E. 90 gear oil. **Radiator:** 10 pints (2 drain taps). **Chassis lubrication:** by grease gun every 500 miles to 9 points. **Ignition timing:** T.D.C. static. **Spark plug gap:** 0.017-0.019 in. **Contact breaker gap:** 0.014-0.016 in. **Valve timing:** I.O., 5° b.t.d.c.; I.C., 45° a.b.d.c.; E.O., 40° b.b.d.c.; E.C., 10° a.t.d.c. **Tappet clearances (hot or cold):** Inlet and exhaust, 0.012 in. **Front wheel toe-in:** 3/32 in. **Camber angle:** nil. **Castor angle:** 3°. **Tyre pressures:** Front 22 lb., rear 22-24 lb. **Brake fluid:** Lockheed Orange (overseas, Lockheed No. 5) **Battery:** 12 volt, 38 amp.hr. **Lamp bulbs:** (Lucas 12 volt). Headlamps, 42/36 watt, No. 354; side and number plate lamps, 6 watt, No. 989; tail/stop lamps, 6/18 watt, No. 361.

Ref.B/9/53

[M]orris Minor (Series II) 4-door Saloon

IN the period since it was announced in the autumn of 1948, the Morris Minor has established a reputation as being a truly outstanding car. Pleasing appearances, roominess within quite compact dimensions, a very moderate fuel consumption, and comfort allied to exceptional controllability on the road have endeared it to innumerable owners—in spite of acceleration which has been quite substantially inferior to that of the majority of other modern cars. Now, as one of the first results of the linking of the Nuffield and Austin organizations into the British Motor Corporation, export four-door saloon examples of the Morris Minor are being built with a power unit closely akin

[ONL]Y SLIGHT re-arrangement of accessories has been necessary to suit the new [engi]ne, which fits snugly in the existing bonnet space and has o.h. valves offering gains [from] both performance and accessibility viewpoints. Note the inclined S.U. carburetter.

A famous small car in new overhead-valve form, with improved performance and undiminished fuel economy

to that of the Austin Seven, and we have recently spent a very pleasant week sampling the notably livelier car thus evolved.

Figures, for once, tell a very great deal of the story concerning the Series II Morris Minor. By contrast with the similar-looking four-door saloon which we reported upon in June, 1951, there is approximately 30 per cent. faster top-gear acceleration all through the speed range, approximately 25 per cent. faster acceleration from rest through the gears, and a maximum speed which is higher by just over 2 m.p.h. Faced with such great performance improvements, it is natural to inquire whether fuel economy has been sacrificed, and in this context it may be recorded that our steady speed tests actually showed economy gains of from 1 to 4 m.p.g., and that our overall consumption figure (which reflects the use made of extra performance) is only ½ m.p.g. below that measured 18 months ago.

Improved performance, it will be noted, results from use of an 803 c.c. overhead-valve engine with 7.2/1 compression ratio, in place of a 919 c.c. side-valve unit with 6.6/1 compression ratio. In its main components, the new engine (and gearbox) is identical to that fitted to the Austin Seven, but one prominent difference is the use of an inclined S.U. carburetter on a different induction pipe; although identical maximum power outputs of 30 b.h.p. at 4,800 r.p.m. are claimed for either engine, performance figures observed on the road indicate appreciably greater torque over a substantial range of moderately high engine speeds when the larger S.U. carburetter is in use.

To obtain extra top-gear acceleration from an engine of smaller swept volume, the Morris Minor top-gear ratio has been lowered, to give 16 per cent. greater engine r.p.m. than hitherto at any road speed. The newer engine, however, has a 15½ per cent. shorter piston stroke, so that theory suggests no loss of cruising speed, and road experience confirms this. A gentle 45 m.p.h. may seem the right cruising speed for a roomy 8-h.p. car, but at a genuine 50 m.p.h. the fuel consumption is still on the desirable side of 40 m.p.g., and whilst a sustained 60 m.p.h. demands almost full throttle it does not increase the fuel consumption to anything worse than 35 m.p.g.

A change which will quite certainly induce divided views is the fitting in conjunction with the new engine of a different four-speed gearbox with slightly wider spacing of the ratios. In effect, whereas the indirect ratios of the former gearbox gave 54 per cent., 130 per cent. and 295 per cent. higher r.p.m. in third, second and first ratios respectively than in top gear, the corresponding percentages for the new gearbox are 68, 159 and 310. It is generally

ATTRACTIVE in its lines, the Minor is in no sense a miniature, its relatively ample width permitting comfortable elbow-room for passengers and also aiding road-holding powers which are well above average.

TALL DRIVERS are well catered for by the adjustable seat, and the body as a whole offers uncrowded room for four people. Four forward-opening doors are a feature of this model.

true to say that the new gear ratios reduce the number of gear changes required in traffic or in hilly country, and an even more notable point is that starting from rest in second gear becomes a normal procedure instead of a slightly naughty one: on the other hand, the keener drivers, who enjoy using pleasant-acting gearboxes such as both old and new ones are, will regret that there is not a faster third gear to improve acceleration when 30-35 m.p.h. traffic is being overtaken. Although not silent when driven up to high speeds in the indirect gears, the power unit does not give any impression of strain, emitting the sort of even "hum" associated with the purely rotary motion of an electric motor rather than the less regular sound expected of a reciprocating engine.

The flexibility of this small engine is very notable indeed, and top-gear pulling remains steady down to 7-8 m.p.h. in top gear. Our test was made on Pool petrol, and towards its conclusion pinking was beginning to be more prominent as carbon formation commenced inside a clean engine—over substantial mileages, the performance figures published on the data page should be maintainable on Premier-grade petrols, but continued use of second-grade spirit might require slightly retarded ignition timing and a 5 per cent. loss of power.

The immediate consequences of installing a new type of engine and gearbox in the existing design of Morris Minor have received first attention in this report, but a certain amount must also be said about the continuing characteristics of the car. The particular point which requires strong emphasis is that, whilst the new 803 c.c. engine is the smallest unit currently fitted to any British car, this saloon is small but does not in any way qualify for the adjective "miniature." Four adults do not feel in any way "crowded up" inside the four-door body, and behind them there is a large luggage locker with external lid.

Exceptional roadworthiness has been a feature of the Morris Minor during its four-year period of production, a number of characteristics combining to give the sort of safe controllability which has not always been associated with 8 h.p. saloons. First and foremost, the rack-and-pinion steering is exceptionally free from lost motion, quite high geared at 2½ turns from lock to lock (with a 33-ft. turning circle), and almost free from the static friction which can destroy steering precision on straight roads. Further, although no car is skid-proof, and this one (with a substantial amount of weight on the front wheels) is quite ready to spin its driving wheels when starting from rest on slippery surfaces, it does seem quite unusually free from vices on greasy roads. There is a certain amount of body roll on corners, but this does not prove objectionable in the absence of sway or tyre howl. The hydraulic four-wheel brakes work inconspicuously well, and the accessible central handbrake would readily lock one rear wheel of the test car on a dry road.

Comfortable Springing

The extreme suspension flexibility productive of what is described in America as the "Boulevard Ride" has been avoided on the Morris Minor, torsion-bar front suspension and semi-elliptic rear springs giving very reasonable shock absorbtion on atrociously pot-holed tracks, but the riding being at its best when the car is maintaining its natural 40-55 m.p.h. cruising speeds. There was a slight impression that perhaps more emphasis had been put upon low-speed comfort on the test car than upon its predecessors, a full load of passengers and luggage failing to disturb the "flat ride," but the rear shock absorbers seeming to allow slightly more spring movement than was necessary under these conditions.

As has been said, this car is dimensioned to take four adults in real comfort, and there is quite ample legroom and headroom in both front and rear. Only the driving seat is adjustable, but a man 6 ft. tall can occupy the fixed front passenger seat without suffering worse than a restricted choice of position. The interior width of the body is, in comparison with pre-war 8 h.p. and 10 h.p. models, immensely generous. One inconspicuous detail which some owners may find most useful is the instantly removable rear-seat cushion and backrest, a few seconds sufficing to make available a huge cargo space extending uninterruptedly from the front seats back to the luggage locker lid. For ordinary "oddments" of which most owners carry all too many, there is quite a roomy locker on the facia, and beneath this a parcel shelf the spaciousness of which was encroached upon by optional-extra radio and heater units fitted to the test car.

Four-door bodywork certainly facilitates access to the rear passenger seats, and for those who prefer to ensure the security of small children a two-door body remains in full production at a lower price. The doors are all front hinged, to minimise any risk of accidental opening when the car is in motion, but entry and exit would be much easier if the front doors could open through a greater angle. It is also unfortunately necessary to record that, in comparison with the two-door model, twice as many doors means twice as many places for draughts to enter the bodywork—a fresh-air heating system to pressurise the body interior might make the simple draught-sealing arrangements adequate, but at present the hinged ventilator panels on the front doors tend rather to lower the car interior pressure at anything above 30 m.p.h. and so strengthen any draughts.

The optional heater, although only of the air-recirculating type, is extremely welcome in wintry weather, and provides a useful amount of warmth in the front compartment quite soon after a cold engine has been started up: perhaps even more important, it de-mists the windscreen really effectively in cool, damp weather. The radio installation, which is another optional extra, suffers from a rather inconveniently positioned control unit adjoining the passenger's knees, but the single downward-facing loudspeaker behind the instruments gave excellent tone and volume.

Already the Morris Minor has made innumerable friends in all parts of the world, and quite evidently the "Series II" Minor with its o.h.v. engine providing more sustained torque is appreciably better than its predecessors. Extra speed and top-gear pulling power, obtained without any sacrifice of fuel economy, make an extremely versatile small car yet more attractive than ever before.

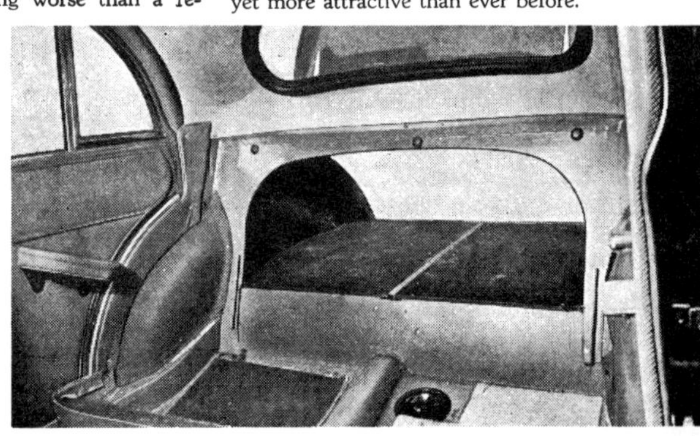

FOR TRANSPORT of bulky articles, the rear seat may be completely removed as shown here, leaving unobstructed room right through to the rear of the boot.

The well-balanced lines of the Morris Minor hold promise of the positive handling for which the car is notable. The bumper over-riders and radio of the test car are optional extras.

ROAD TESTS FROM THE OWNER-DRIVER ANGLE—

No. 9
THE SERIES II MORRIS MINOR

HALF a dozen people who drove a Series II Morris Minor which was in our hands recently expressed their delight with the car and all had a different reason for so doing. The unexpected roominess, the steering, the road-holding, the liveliness, the top-gear flexibility, all earned their share of praise from drivers—and passengers, too—whose motoring requirements were by no means identical.

The Series II saloon is amongst the first fruits of the British Motor Corporation which combines the manufacturing resources of the Austin Motor Co. Ltd. and Morris Motors Ltd. It differs from the original Minor principally in its power unit, an overhead-valve engine of 800 c.c. having replaced the original 919-c.c. side-valve model, and resulting in improved performance without loss of fuel economy.

This latest engine is identical in its general characteristics with that installed in the Austin A30 " Seven " (a road test report of which appeared in *The Light Car* of April, 1953). One important difference concerns carburation: the Morris Minor is equipped with an inclined S.U. instrument.

Thousands of drivers are already aware of the qualities of the Issigonis-designed Minor which was first introduced in the autumn of 1948. Its shapely, well-balanced lines, firm yet comfortable suspension and positive steering all earned for the car an enthusiasm such as is more usually evoked by a sports car. Many owners voiced the opinion that such road-holding deserved more power: indeed, there are twin-carburetter and supercharged versions on the road which give altogether surprising performance figures and much innocent amusement to their conductors.

Clearly in sympathy with this view the manufacturers have now produced the power where it is most useful—at the lower end of the speed range—without losing sight of the fact that the Minor is intended to be a small family car of economical habits.

Interior trim is pleasingly practical. The instruments are grouped in front of the driver, hand-brake and gear lever are centrally placed and the individual seats give good support.

The maximum speed of the Series II is no more than a mile or two an hour faster than its predecessor, but its acceleration from rest, and in top gear at speeds in excess of 40 m.p.h., has been stepped up by as much as 25 per cent. It might reasonably be supposed that this improved performance would be offset by an equal increase in fuel consumption. This is by no means the case: at certain speeds the new engine is as much as 4 m.p.g. more economical and in overall running, including long periods of sustained high-speed cruising and town work, the figure of 39 m.p.g. is only fractionally higher than that obtained with the side-valve engine.

Different gear ratios, more widely spaced and lower on top, contribute in large measure to the liveliness of the new Minor. Fourth gear is now 5.28 to 1 and third 8.88, the latter perhaps a little low for the man who enjoys a fast middle ratio. Valve bounce sets in at about 45 m.p.h. in third gear but in practice we found that to change into top at 40 m.p.h. produced an equally good figure for the " flying quarter-mile," for top-gear acceleration between 40 and 60 m.p.h. is smooth, rapid and consistent.

In spite of its small capacity, the remarkable flexibility of the engine makes the Minor very nearly a top-gear car. It will pull away from 8 m.p.h. without effort, quickly reach its cruising speed of 50-55 m.p.h., and hold that speed on normal main road hills. Throughout its range engine noise is unobtrusive but the intermediate gears do not reach the same standard of silence.

The Minor is one of those cars in which one is quickly at home in the driving seat. There is ample elbow room, the steering wheel is comfortably raked, the pedals are well spaced and the instruments are where they should be—in front of the driver. The gear lever projects directly from the gearbox and the handbrake is located out of the way between the individual seats. In spite of rather stout screen pillars forward visibility over the short bonnet is good.

In the cool dry weather during which we tested the car only the briefest use of the choke was necessary. The engine warms up quickly and even after a night in the open is soon developing its full power. Pedal pressures are light, gear-changing rapid and foolproof and the brakes are effective throughout the speed range with little or no appreciable fade after repeated applications.

The interior layout of the Series II saloon is pleasing and practical. The seats, the backs of which can be tilted forwards, are perhaps somewhat firm and upright, but nevertheless give a comfortable ride over long distances. Only the driver's seat is adjustable, but adequate leg room removes any possibility of being cramped. At the rear the bench-type seat accommodates two adults in comfort with plenty of headroom, good vision through the windows and ample foot space beneath the front seats.

The luggage boot is surprisingly capacious (the spare wheel is carried on a separate lower shelf) and an unusual and useful feature is the removable back-seat squab, the absence of which converts the Minor into a small station-wagon. Small parcels can also be carried on a shelf under the facia panel, in a glove box opposite the passenger and on a narrow shelf beneath the rear window.

Driving the Morris Minor is among the most pleasurable of motoring experiences. The impression of good balance suggested by the lines of the car is confirmed by its handling on the road. There is a firmness about the suspension which imparts complete confidence at any speed and the rack and pinion steering is light and free from wander-producing lost motion. The car runs absolutely straight over the worst surfaces with only finger-light control and can be cornered with a zest more normally reserved for the sports car.

Its compact size (in conjunction with a gearbox which the enthusiastic driver may well wish to use to the full) makes the Minor ideal transport for dealing with town and arterial road traffic. Even more impressive, however, is its performance over winding secondary roads. Unaffected by poor surfaces, camber and anything but a major gradient, the Series II will maintain average speeds worthy of a much more powerful car and moreover without disconcerting driver or passengers for a moment. Under any conditions it will scurry along, instantly responsive to the technique of the driver, without fuss, fatigue or fits of temperament.

The Light Car
Made-to-Measure Profile Diagram

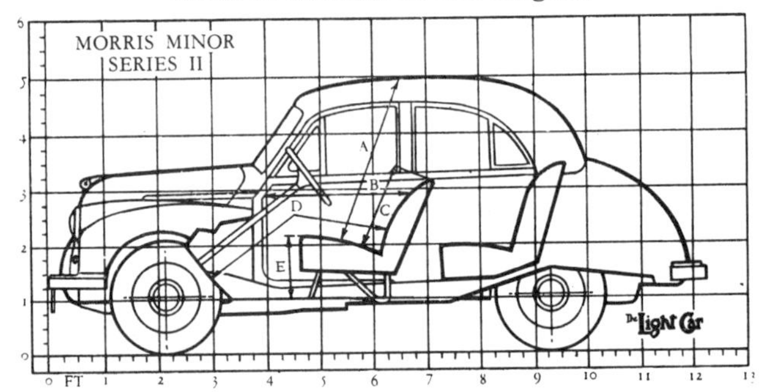

Key dimensions: A, 36½ in., B, 14¼ in. max., 11 in. min., C, 19 in., D, 40¼ in. max., 36¾ in. min., E, 12½ in.

SPECIFICATION IN BRIEF

Engine.—4-cyl., o.h.v., 58 mm. by 76 mm. (800 c.c.), 30 b.h.p. at 4,800 r.p.m. Inclined S.U. carburetter, S.U. electrical petrol pump. Sump capacity, 6¾ pints (3.8 litres). Sparking plugs, Champion NA.8.

Transmission.—Borg and Beck single dry plate clutch: gears, 5.286, 8.88, 13.69 and 21.618 to 1. Reverse, 27.38 to 1. Rear axle, hypoid bevel.

General.—Suspension: front, independent (torsion bar and wishbone), rear, semi-elliptic leaf. Shock absorbers, front and rear, Armstrong hydraulic. Brakes, Lockheed 7-in. dia. hydraulic. Tyres, Dunlop, 5.00 by 14. Electrical equipment, Lucas, 12-volt, 38 amp. hr. Fuel tank, 5 galls. (22.17 litres).

Dimensions (see chart).—Width, 5 ft. 1 in.; wheelbase, 7 ft. 2 in.; track, front, 4 ft. 2⅝ in., rear, 4 ft. 2 5/16 in.; clearance, 6¾ in. Turning circle, 33 ft. Unladen weight, 15¾ cwt.

Price.—£405, plus £169 17s. 6d. Purchase Tax (£574 17s. 6d.).

Manufacturers.—Morris Motors, Ltd., Cowley, Oxford.

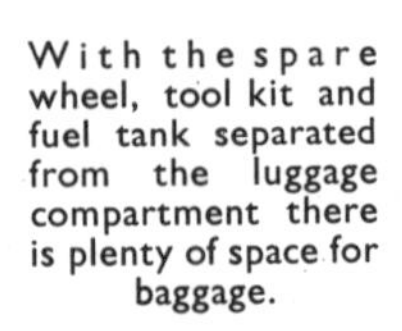

With the spare wheel, tool kit and fuel tank separated from the luggage compartment there is plenty of space for baggage.

Servicing charges today form a large part of the overall cost of running a car. On many current models only modern lubrication equipment can reach the numerous points which require regular attention. The owner-driver of the Minor, however, can do much of this kind of routine maintenance work himself with the minimum of effort and with a considerable saving in cost.

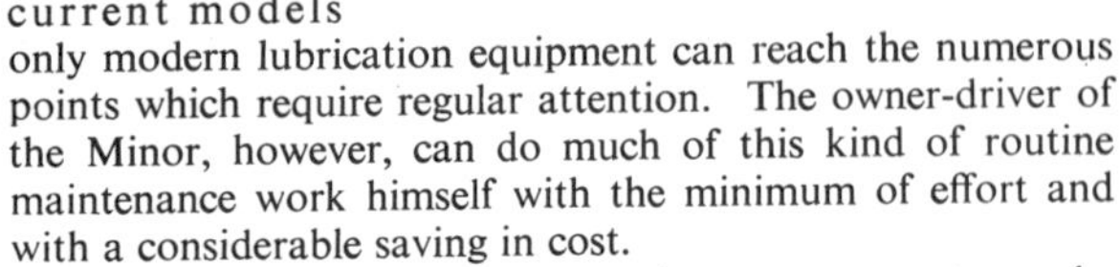

The whole of the front suspension arrangement can be reached without difficulty either with the bonnet raised or from under the car. Carburetter, ignition components and the front shock absorbers are all unobstructed, inviting the regular attention they deserve. Moreover, its compact overall size and smooth contours make the Morris coachwork amongst the easiest to wash down.

Here is a car which admirably reflects one's driving moods. It will pull smoothly and powerfully in top gear in traffic or potter quietly through the lanes, virtually ignoring the gradients. In brisker company it will romp along, zestfully putting the miles behind, showing no fatigue—and little thirst for fuel. Whether it is driven with a full complement of parents and children, or by an enthusiast alone, this little car appears to do its work with a willingness which is a constant source of pleasure to the driver.

Although the car submitted for test was a four-door model, all Minors are now available with the o.h.v. power unit. An external feature which distinguishes these models from the side-valve cars is the additional bonnet motif.

The Morris Minor is by no means merely an elegant small town carriage: it is a touring car with the traditional British characteristics of sturdiness, safety and economy.

The compact layout of the 800-c.c. engine: carburetter, electrical auxiliaries and even the shock absorbers are immediately accessible. The inclined S.U. carburetter is a distinctive feature of the o.h.v. 800-c.c. Morris engine.

ROAD AUTO AGE TEST

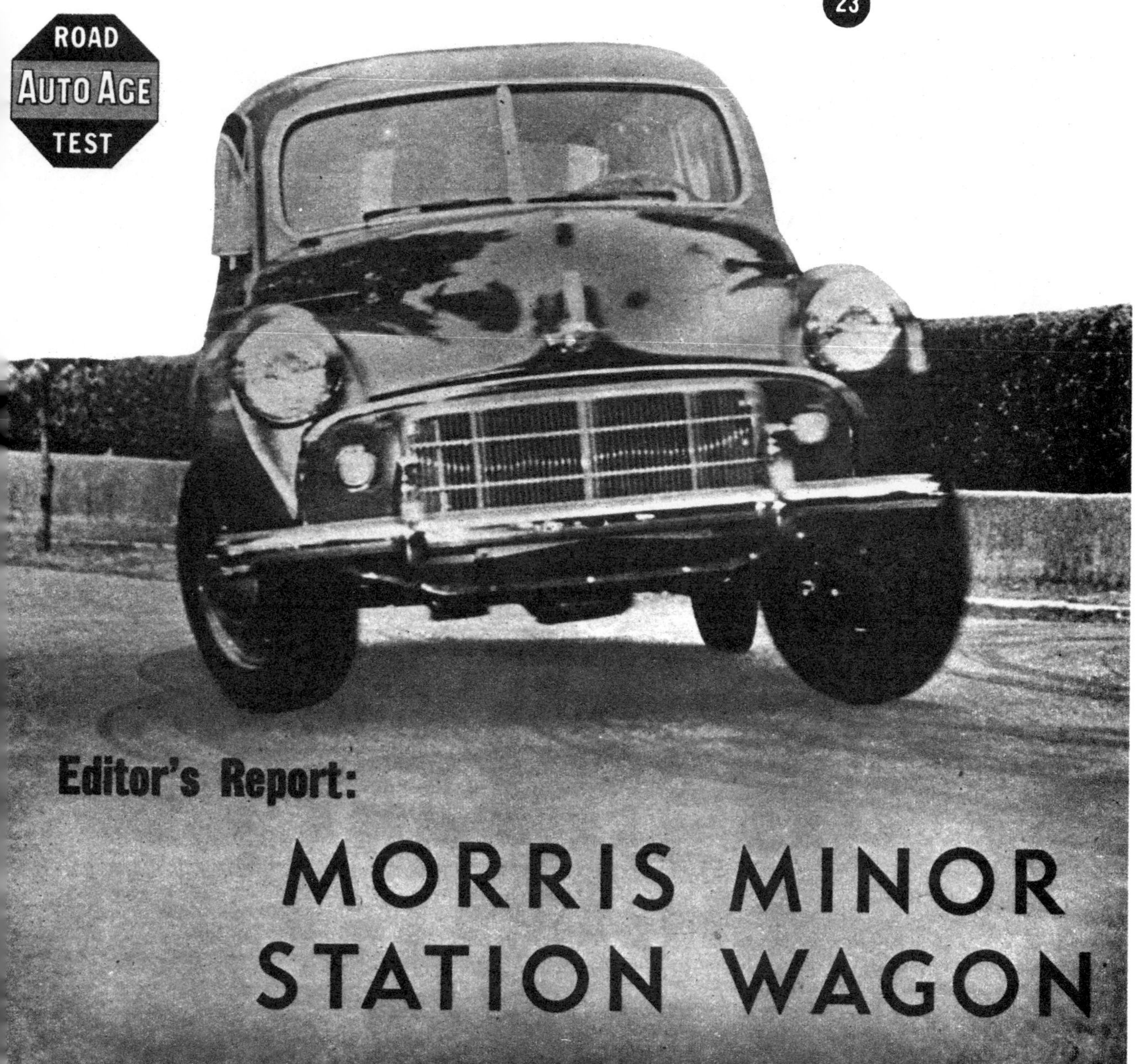

Editor's Report:

MORRIS MINOR STATION WAGON

PHOTOS BY DAN RUBIN

Here the author gives the Minor a last-minute check before taking-off on a trial spin. Note that while the car is low it is not squat. There is plenty of head room; no need to be an acrobat to get in and out.

By HARVEY B. JANES

Want a station wagon that gets 40 miles to a gallon, parks on a nickel and handles like a sports car? Try the Morris.

In profile the Morris station wagon looks surprisingly graceful. It is constructed mostly of metal, but that's real wood in the rear section.

IF I were to come right out and state flatly that all station wagons handle like slobs, this office would no doubt be flooded with hundreds, if not thousands of letters from irate station-wagon owners demanding my scalp and a retraction—or both. Now I freely admit that even to discuss a station wagon on the basis of handling and cornering is, for the most part, completely unfair to a vehicle that is intended mainly as a heavy-duty carryall with acres and acres of space for passengers and assorted freight. Still, I think that most of you will agree that the average station wagon is somewhat less stable on the road than a comparable sedan model. Get what I'm driving at? Fine. Now enter the Morris Minor.

At this point let me tell you quickly about the ordinary Morris Minor sedans and convertibles, in case you don't know. They are, of course, little brothers to the famed MG and are powered, or have been since 1953, by neat little 800-cc, overhead-valve engines identical to the ones used in the Austin A-30. This gives them a top speed of near 70 mph and the ability to cruise at 60 all day long. Not terribly thrilling? Not in a straight line, no, but when you get one of them on a narrow winding road, especially downhill, hang onto your hat, because these babies will actually outhandle MGs, Jags and practically any other sports car you can name. Why? Superb suspension is the answer, coupled with proper weight distribution, fast rack-and-pinion steering and a real tiny 86-in. wheelbase.

I was aware of all this as I stepped out of the taxi in front of J. S. Inskip's in New York, just prior to picking up the Morris station wagon I was to test over the weekend. Still, I had my doubts, and they weren't lessened any when I saw the small, gray car parked at the curb. Sure, it was a cute-looking little monster with very nice

Huge air filter over the single SU carburetor practically hides the 30-hp, 800-cc overhead-valve engine. The electric fuel pump, extreme right, is too close to exhaust manifold; this tends to cause vapor lock.

Also inside the engine compartment, on the left and right-hand corners just forward of the fire wall we find, of all things, the front shock absorbers. These permit easy servicing without getting under the car.

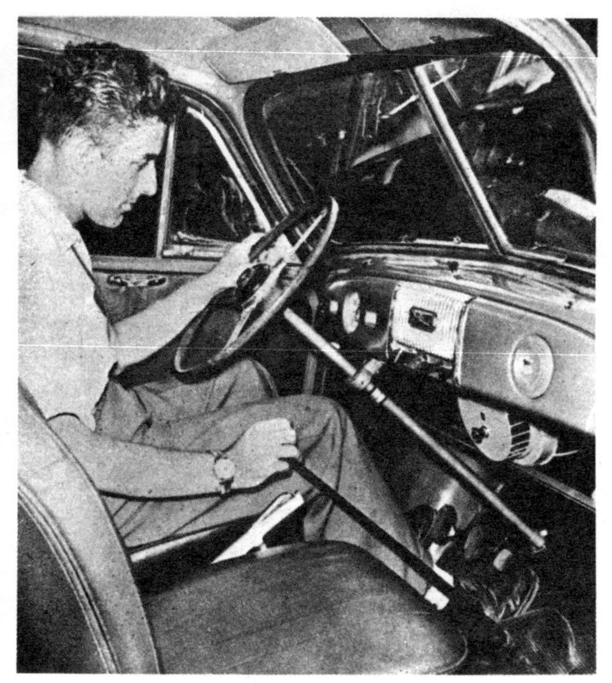

Individual "chair height" bucket seats are leather upholstered, extremely comfortable. Clutch, brake and gas pedals are placed rather close together, still work smoothly with heels resting on the floor.

With the rear doors open and the back seat folded flat the Morris is an amazingly roomy car, considering its 12-ft., five-in. over-all length. Spare tire and tool kit are housed in a separate compartment below.

appointments, as the British would say, and a surprising amount of room. But it looked a lot bigger than the Minor sedan and it had to be heavier. The question was, how would it perform?

There was only one thing to do; I climbed behind the wheel. So far, very nice. The twin bucket seats in front were neatly upholstered in real leather and were, in fact, a bit more heavily padded than those on the Morris sedans and convertible. They also seemed to be set somewhat higher off the floor, thus getting away from the legs-out-in-front-of-you effect usually found in small foreign cars. You may or may not like this chair-height feature but it does give you more fore-and-aft room, a good point in a station wagon, especially a small one.

My first shock came when I glanced at the odometer; it read 23 miles, hardly conducive to a real rough-and-tumble test. This was a brand new car which meant that I had to hold it down to something like 12 mph in first gear, 18 in second, 26 or 28 in third and 45 in high. Even that was stretching it. Oh well, I thought, maybe I won't want to push it anyway. I started it up, goosed it a few times to get the feel of the pedal and then eased away from the curb. And I mean eased. Two small boys, obviously intrigued by the "cute little car," raced me to the traffic light at the corner. They won.

The light had changed to red. While I waited for it to turn back to green, one of the men from Inskip's pulled up alongside me in an MG.

"Drive it like it had an automatic transmission," he called up to me.

"What the hell do you mean?" I asked politely.

"Start her off in third gear," he told me.

Of course I thought he was out of his mind, but when the light changed I slipped it into third and let the clutch out dubiously, feeding lots of gas. The Morris moved away surely and smoothly, without a trace of a buck. No acceleration to speak of, naturally, but when you think of an engine displacing just slightly over three-quarters of a liter pulling more than 2,000 lbs. (the combined weight of me and the car) from a standing start in third, you realize that this feat was nothing short of amazing. Later on I tried the same thing with four adults in the car with the same results. The Series II Morris Minor—the one with the overhead-valve engine—was designed with the American woman in mind, very definitely.

By the time I reached home I found that I was completely at ease in the car. And I had found out a few other things, too. This was one of the hottest days of the year and I had been forced to crawl along for almost an hour through heavy stop-and-start traffic. Scores of cars, old and new, had overheated on all sides of me, but the Morris gave no complaint at all except that its SU electric fuel pump, which is placed too near the hot exhaust manifold, began to tick away like an insomniac's dollar-and-a-half alarm clock—a sure sign of vapor lock. Still, the car never missed a beat.

I parked the Morris in front of my house, flew upstairs to consume a dinner in record time and dashed downstairs again. Sure enough, the car was still there; none of the neighborhood children had thought to pick it up and carry it away. Delighted, I climbed behind the wheel, curious to find the answers to a number of questions that the car had forced into my head on the ride home from Inskip's. I hadn't had any chance to test it with all the traffic, but the Morris had seemed remarkably stable to me, station wagon or not. After promising myself that I would hold the revs down to about 3,200, I started it up again and headed for a favorite "test" grounds near my home.

With the tires checked at an even 24 psi, I started getting a little frisky. A few right-angle turns at 25 mph produced a little tire squeal but hardly any roll and absolutely no sway at all. So I pushed the speed up to 30 mph. Still no trouble. Then I got a bright idea. Cruising down a deserted street at 30 mph, I suddenly cut the wheel hard to the left and the little car executed a perfect U-turn, with the rear end following like a dirt-track midget. And I had done it in high gear, mainly because I didn't want to rev the engine high enough to get it

Taking a hard right-angle corner at better than 40 mph the Morris squeals a bit and makes distinct tire marks on the ground, but simply refuses to break loose into a spin. Note almost complete lack of lean.

With all the tests completed the author gives the tires a sympathetic pat for a job well done. Standard 5.00 x 14 Dunlops are fitted on all Minors giving many miles of service, good traction on all surfaces.

into third at 30. This was fun. I drove back down the street and tried it again at 35 mph. Same thing, except this time I caught a distinct odor of Dunlop No. 5 drifting on the warm summer air. What the heck, I thought. What are tires to Inskip? So I backed off way down the street and tried it again at 40, and here I got the biggest surprise of all. At that speed, with the additional revs, I had more horsepower, and the Morris made the 180-degree turn even better than it had at 30—if you could call it a turn. What it actually did was to spin around right on a spot and then head back the other way. The Dunlops didn't like it one bit but I enjoyed it thoroughly. I spent the rest of the evening tooling the little job in and out the streets of the Bronx and Queens, trying to put on mileage to loosen up the engine. It was feeling better all the time and so was I.

On Saturday I took the car up into Westchester and tried it out on some of my favorite roads in that area. By this time I was thoroughly delighted with it except for a few minor things. (No pun intended.) In the first place, holding it down to the 45 mph break-in limit on the open road was a real drag; people were passing me with the dirtiest looks. Secondly, Saturday turned out to be *the* hottest day of the year—almost 97 degrees—and although the Morris showed no signs of heating up, I still felt the lack of a water temperature gauge. It's nice to know how close you're coming to trouble. The only other thing that bothered me at all was the gearbox—pardon me, Percy—transmission. Now don't get me wrong—it's not a dog or anything. In fact, the synchromesh is real good and the gearbox is actually one of the easiest and fastest shifting ones you'll come across. It's just that I don't happen to like the choice of gear ratios; first is too low and third doesn't have enough speed on it. I'm told that if you get the engine really broken in correctly, you can squeeze 49 mph out of the thing in third, but even that is sort of slow. Of course, I'm one of those crazy guys who's not happy unless he's shifting gears all the time, I like to control it myself; sue me. For that reason I still prefer the gearbox on the old flat-head Morris. There goes progress down the drain.

By Saturday evening I had put about 125 miles on the Minor and I decided to fill the tank again to get an idea of the gas mileage. It figured out to be about 25 mpg, which is not bad at all, considering the newness of the car and the type of driving I had been doing. Now I wanted to take it out on a *Continued*

SPECIFICATIONS

ENGINE: four-cylinder, overhead valves; bore, 2.28 in.; stroke, 3.00 in.; total displacement, 49 cu. in. (800 c.c.); developed hp, 30 at 4,800 rpm; compression ratio, 72 to 1; maximum torque, 40 lb./ft. at 2,400 rpm; carburetor, single SU; ignition, 12-volt.

TRANSMISSION: four-speed manual with synchromesh on top, third and second gears.

REAR AXLE RATIO: 5.286 to 1.

SUSPENSION: front; individual torsion bars and links; rear, semi-elliptic leaf springs.

BRAKES: hydraulic, with two leading shoes in front, one leading and one trailing shoe in the rear.

DIMENSIONS: wheelbase, 86 in.; track, 50⅝ in. in front, 50 5/16 in. in the rear; width, 5 ft., one in.; height, 5 ft., ½ in.; over-all length 12 ft., 5 in.; ground clearance, 6¾ in.; turning circle, 33 ft. 11 in.; dry weight, 1,848 lbs.; tires 5.00 x 14.

PERFORMANCE

ACCELERATION: 0 to 30 mph: 7.9 seconds
0 to 50 mph: 24.2 seconds

TOP SPEED: 67-69 mph (approx.)

Morris Minor Station Wagon

good long run, so I decided to drive out to Montauk Point at the tip of Long Island. You know—heat wave and all that? Besides, I was thinking of some great roads. . . .

Early Sunday morning I shoved off, creeping out along the Long Island highways, religiously holding the Morris to a lousy 45 mph, getting madder all the time. About an hour-and-a-half out, I stopped at a friend's house and took him and his wife and kids for a short ride, sort of curious what the family-type reaction would be. He and his wife seemed impressed with all the usable space in the car and the kids just loved the whole thing. But you know children with any small car, or just any new car, for that matter. Like a new toy.

Twenty minutes later I was back on the road and now the traffic was starting to thin out as I put the miles behind me. The tempo picked up and I was really having a battle on my hands holding the speed down. I was losing the battle. Time and time again I found that little red needle climbing up to 50 and better but the car didn't seem to mind, so I let it go. The oil pressure gauge stayed at about 55 all the way; every once in a while the speedo joined it. For a while I thought the gas gauge was broken because it wasn't moving at all. I used to drive a gas-gobbling classic Lincoln; the loudest noise it made was glub, glub.

I breezed into the town of Montauk at about 1:30, parked the station wagon and ate a leisurely picnic-type lunch. The car sure attracted attention. People from eight to 80 kept stopping to ask questions and I didn't mind at all because every time I showed the Morris off to someone, I learned something new about it myself. What people mostly wanted to know was how fast would it go, what kind of gas mileage did it get and how much junk would it hold. I guessed at the answers to the first and last questions, but after a while I decided to get a definite figure for the one about mileage. The odometer now read 281, so I subtracted and found that I had run 156 miles since the tank was last filled. I managed to squeeze 4.3 gallons of go-juice into the slightly-over-six-gallon tank, which figured out to a little better than 36.3 miles per gallon. This was more like it. My guess now is that after four or five thousand miles the Morris should be able to get better than 40 miles on a gallon of gas at a steady speed of 40 mph or so. Okay?

With the questions of the entire town answered, I hopped back into the wagon and started on a grand tour of the Point and surrounding countryside. Real beautiful stuff that Montauk.

It didn't take me long to get tired of that straight old Montauk highway, though, so I headed off onto some of the more interesting side roads. Still much too smooth. I wanted to find out how the darn thing rode on really bad stuff, so I found a little washboard cinder path alongside the tracks at the railroad station and let her fly. Funny thing about these firm-suspension cars. They'll bounce like mad on a second-class road, especially if they're not loaded, but when the going really gets rough, they begin to smooth out again. The faster I went on the cinders the better the Morris liked it. Once again the Dunlops must have hated me just a bit but even they didn't actually complain at all. Convinced and dirty, I got back onto the blacktop again. To tell the truth, the station wagon does ride sort of hard by our standards, especially in the rear, but they have to beef it up back there with heavy-duty semi-elliptic leaf springs so the car won't sag when it's heavily loaded as any station wagon is apt to be. Just as a matter of comparison, it always rides much softer than an MG.

When I let it lead me on, the blacktop started to get interesting—very interesting indeed. I found myself on a tight, narrow, winding road, climbing sharply up and down hill all the way. If this wasn't a place to test the roadability and handling characteristics of a car, I never saw one. Feeling my way, keeping the car in high, I began to increase my speed to 35, then 40, then 50 mph. Man, this was great! That station wagon went up and down and around corners like a gone-wild roller-coaster, drifting just enough whenever necessary, with little more than a hint and a nudge from me. It stayed flat all the way, steered like a dream, and before I realized it I was running my own private Montauk Grand Prix, zipping along at nearly 60 mph on roads that certainly were never meant for speeds like that. No matter how hard I tried, I couldn't seem to break the rear end loose into anything more than a nice, slow, controlled drift. So help me, this is the only station wagon I have ever been in that handled just like a sports car.

Now this may not be a selling point to you. You are probably still asking, and with good reason, just how the Minor would rate as a station wagon in the usual sense of the word. For answer I can only say that it all depends on what use you have for a car. If you want a station wagon that you can kick around, one that will carry nine people to and from the train and then turn around and haul a few hundred pounds of sand and cement and 12-ft. pieces of lumber, just forget the Morris. It's not the car for you. And if that's what you want a car for, you probably won't care whether or not it corners too well anyway. But if you're looking for a buggy that's fun to drive, and safe, and can carry huge amounts of groceries, hardware, luggage or what-have-you, then it's something to consider seriously. It's not terribly fast—in fact, I saw no point in running off any real performance tests on it, aside from a few simple stopwatch checks. It wouldn't have proved anything anyway; the car was too new. On the other hand, that gas economy is nothing to sneer at, and the Morris station wagon does hold more luggage than any small car I have seen—actually more than any of our big sedans. Its space is usable, too, and plenty easy to get at with those pick-up truck rear doors.

It seems to me that in view of the easy handling, light steering, and good, positive brakes, this is just a perfect car for a woman. For shopping in and around town, parking in small spaces or winding along narrow country lanes, it just can't be beaten. Even slippery roads in winter will hold no terror for this baby. Most women will like the way it starts in second or third and the way it pulls like mad on hills in high gear. They won't have any trouble getting it into the garage, either, because believe it or not, it measures just one inch longer than the regular Minor sedan and is exactly the same width. And the old man may even borrow it for a camping trip now and then when he finds that he can sleep in it with the rear seat folded flat. I did; real comfortable.●

The Motor Road Test No. 29/56

Make: Morris. **Type:** Minor 1000 de luxe 4-door saloon

Makers: Morris Motors Ltd., Cowley, Oxford

Test Data

CONDITIONS: *Weather: Cool, misty weather with light wind. (Temperature 44°—45°F., Barometer 30.2 in Hg.) Surface: Smooth tarred. Fuel: Premium grade, approx. 95 Research Method Octane Rating.*

INSTRUMENTS

Speedometer at 30 m.p.h.	1% slow
Speedometer at 60 m.p.h.	accurate
Distance recorder	1% fast

WEIGHT

Kerb weight (unladen, but with oil, coolant and fuel for approx. 50 miles)	15¼ cwt.
Front/rear distribution of kerb weight	57/43
Weight laden as tested	19 cwt.

MAXIMUM SPEEDS

Flying Quarter Mile

Mean of four opposite runs ..	.. 72.4 m.p.h.
Best one-way time equals ..	.. 74.4 m.p.h.

"Maximile" Speed (Timed quarter mile after one mile accelerating from rest)

Mean of four opposite runs ..	.. 70.9 m.p.h.
Best one-way time equals ..	.. 73.2 m.p.h.

Speed in Gears

Max. speed in 3rd gear	61 m.p.h.
Max. speed in 2nd gear	36 m.p.h.
Max. speed in 1st gear	23 m.p.h.

FUEL CONSUMPTION

55.0 m.p.g. at constant 30 m.p.h. on level.
54.0 m.p.g. at constant 40 m.p.h. on level.
46.5 m.p.g. at constant 50 m.p.h. on level.
40.0 m.p.g. at constant 60 m.p.h. on level.

Overall Fuel Consumption for 991 miles, 27.3 gallons, equals 36.3 m.p.g. (7.8 litres/100 km.)

Touring Fuel Consumption (m.p.g. at steady speed midway between 30 m.p.h. and maximum, less 5% allowance for acceleration). 42.9 m.p.g.

Fuel Tank Capacity (maker's figure) 5 gallons.

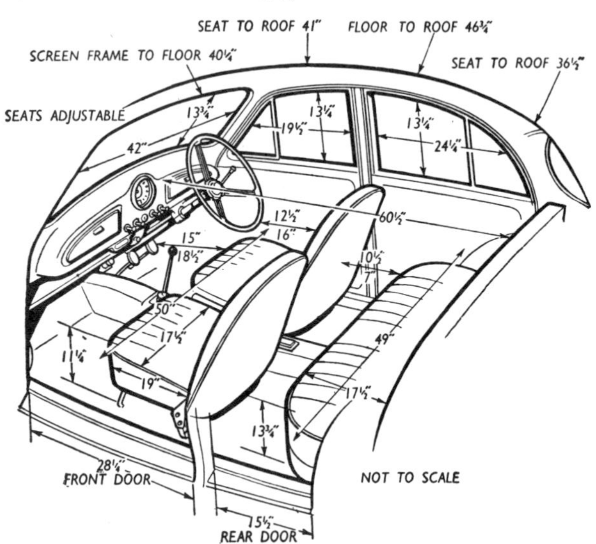

ACCELERATION TIMES from standstill

0-30 m.p.h.	6.9 sec.
0-40 m.p.h.	12.3 sec.
0-50 m.p.h.	18.7 sec.
0-60 m.p.h.	30.1 sec.
Standing quarter mile	24.1 sec.

ACCELERATION TIMES On Upper Ratios

	Top	3rd
10-30 m.p.h. ..	14.2 sec.	8.9 sec.
20-40 m.p.h. ..	14.9 sec.	9.7 sec.
30-50 m.p.h. ..	17.1 sec.	12.1 sec.
40-60 m.p.h. ..	22.7 sec.	23.0 sec.

STEERING

Turning circle between kerbs:

Left	34¾ feet
Right	30 feet
Turns of steering wheel from lock to lock	2½

BRAKES from 30 m.p.h.

0.95g retardation (equivalent to 31½ ft. stopping distance) with 90 lb. pedal pressure
0.80g retardation (equivalent to 37½ ft. stopping distance) with 75 lb. pedal pressure
0.56g retardation (equivalent to 54 ft. stopping distance) with 50 lb. pedal pressure
0.32g retardation (equivalent to 94 ft. stopping distance) with 25 lb. pedal pressure

HILL CLIMBING at sustained steady speeds

Max. gradient on top gear	.. 1 in 12.8 (Tapley 175 lb./ton)
Max. gradient on 3rd gear	.. 1 in 8.6 (Tapley 260 lb./ton)
Max. gradient on 2nd gear	.. 1 in 5.4 (Tapley 410 lb./ton)

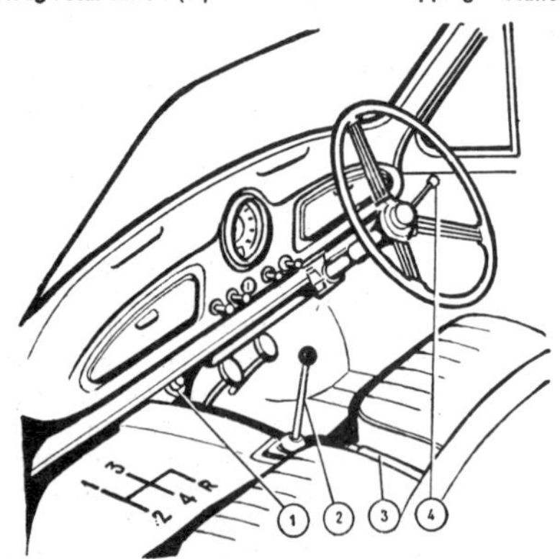

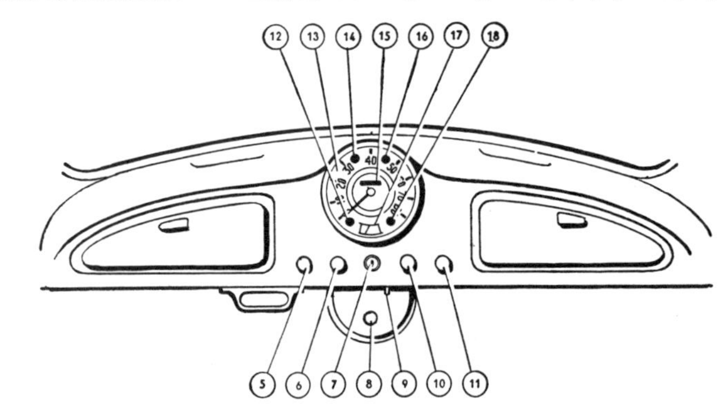

1, Headlamp dip switch. 2, Gear lever. 3, Handbrake. 4, Horn and direction indicator control. 5, Choke control. 6, Windscreen wiper control. 7, Ignition switch. 8, Heater fan switch. 9, Panel light switch. 10, Head and sidelight switch. 11, Starter control. 12, Oil pressure warning light. 13, Speedometer. 14, Direction indicator repeater light. 15, Distance recorder. 16, Headlamp main beam indicator light. 17, Fuel contents gauge. 18, Dynamo charge warning light.

The MORRIS Minor 1000 de luxe four-door saloon

An Old Favourite with a Stimulating New Performance

From the front the principal change to be seen is a curved glass one-piece windscreen which replaces the V-type used on all earlier models of the series. The direction indicators on the four-door model are mounted high in the centre pillar. On the two-door version they are located below the waistline.

THERE has never been a bad Morris Minor. Right from the beginning of the marque in 1930 the intention would seem to have been to produce a thoroughly well-made small car with a high standard of finish and a degree of performance calculated to satisfy the hard driver over a long period. At the same time the Minor has retained a rugged reliability and ease of operation which has endeared it to families all over the world.

The post-war Morris Minor introduced in 1948 was entirely new from stem to stern, and in fact was so successful that it placed Britain well ahead of all her rivals in one of the most competitive markets in the world. It is a tribute to the Minor that despite the most advanced creations from Italy, France and Germany the sales of the Morris remained so strong that only recently did it become necessary to carry out any serious modifications to the car as originally conceived.

For 1957 the Nuffield engineers have produced an outstanding little vehicle. The latest engine develops 23% more power than the old, and the cubic capacity has gone up from 803 c.c. to 948 c.c. This increase is achieved by a new cylinder block with siamesed bores which have been increased from 58 mm. to 62.94 mm. The compression ratio has been raised from 7.2 : 1 to 8.3 : 1, and there is a new distributor. To allow for the extra power the diameter of the big-end bearings has been increased and lead-indium is being used as a lining material. This alloy was one of the factors which enabled the prototypes of the new car to average over 60 m.p.h. for 25,000 miles in Germany earlier this year and others were undoubtedly the stronger connecting rods introduced on the new car together with a more robust crankshaft.

To cope with the increase in power the clutch has been strengthened, and the final drive ratio is raised to permit higher cruising speeds at less r.p.m.

These then are the hidden improvements which go into the new Morris Minor. Externally there are ways in which the car differs slightly from its well-known predecessor. The vee front windscreen has been replaced by a curved single sheet and

In Brief

Price: £445 plus purchase tax £223 17s. 0d. equals £668 17s. 0d.

Capacity	948 c.c.
Unladen kerb weight ...	15½ cwt.
Acceleration:	
20-40 m.p.h. in top gear ...	14.9 sec.
0-50 m.p.h. through gears	18.7 sec.
Maximum direct top gear gradient	1 in 12.8
Maximum speed	72.4 m.p.h.
Maximile speed	70.9 m.p.h.
Touring fuel consumption ...	42.9 m.p.g.

Gearing: 15.18 m.p.h. in top gear at 1,000 r.p.m.; 30.5 m.p.h. at 1,000 ft./min. piston speed.

The short, stiff, almost sports-car type of gear lever is particularly pleasant to use. Other worthwhile features which can be seen in this photograph include the sensibly located hand-brake, a remarkably clear speedometer combined with fuel gauge and four tell-tale lamps, two lockers above the full-length parcel tray and the well positioned pedals.

of the parcel tray and was of the recirculation type without access to its own fresh air supply. As a heater and de-mister it was undoubtedly successful but books on the parcel shelf were inclined to curl their edges from the warmth of the blast and it would certainly be no place for the family shopping if butter or chocolate were involved. Draught-free ventilation could be obtained only by closing the front windows and panels and opening both rear windows about half an inch.

Further small parcel space occurs behind the rear passenger seat where there is a useful shelf. Its metal base, however, can act as a most mystifying sounding-board, and in the interests of peace and quiet soft goods only should be placed thereon.

the size of the rear window has gone up substantially. The words "Minor 1000" on the boot lid identify the new car while internally a redesigned facia panel may be observed.

So much for the background and development of the latest Minor. A journey of 212 miles which commenced at 5 a.m. proved the most interesting introduction to the newly acquired virtues of the little car. It also revealed certain characteristics which will be criticized, but in balance the effect was to engender a great deal of enthusiasm and a very real affection for the new model.

The car which we drove possessed four doors, and we did not think that these opened sufficiently widely to make entry and exit easy for the portly. The driving seat which is adjustable with effort is reasonably comfortable, although the range of rearward travel could be increased with advantage by at least another inch in order to house the exceptionally tall driver. All-round visibility is now commendable and the short, stiff sports-car type gear lever a real pleasure to use. The handbrake lies snugly between the seats in an almost perfect position and the pedals have just the right rake and angle. The dipper switch is easily found, which is a most important and rather rare attribute, but the switch itself failed as soon as we used it. The fact that it was possible to obtain and fit a replacement for the offending device within an hour at 9 o'clock at night provided an unexpected confirmation of the soundness of the Nuffield spares organization.

The direction indicators, which are of the illuminated arm type, functioned well throughout the test at all speeds but they are actuated by a control which must be censured. This little lever, which extends below the steering wheel, has no self-centring action and the responsibility of reminding the driver to replace the indicator after use rests with a pale green light almost unnoticeable in daylight and totally unnecessary at night. Moreover, the same lever must be pressed in to sound the horn, and were it possible to do this with a light touch of the finger some justification for the scheme might exist. As, however, it seems necessary to remove the hand from the wheel in order to bring enough pressure to bear, any advantages which might have been gained from intelligent positioning are lost.

The lights of the car are fully in keeping with the new performance, and enable high average cruising speeds to be maintained even in poor conditions at night. Two small lockers with ingenious snap-fitting doors occupy both sides of the facia panel, and there is a full-length parcel tray a little lower down. The heater fitted to the car which we tried was mounted in the centre

From the rear the new model is distinguished by the words "Minor 1000" and an enlarged, curved rear window. The lid of the boot may be opened right up so that it rests against the top beading of the rear window. The spare wheel and tools are separated from the luggage by a flat platform.

The keen motorist will appreciate that these afore-mentioned criticisms are of no great import and the remedy for several of them can most easily be discovered. Among the many considerable attractions of the Minor, however, must be listed a most intelligent rooflight directly over the top of the windscreen; a degree of air stream and mechanical silence at all speeds which is highly satisfactory and a pair of self-parking windscreen wipers which not only do their work in torrential rain but seem capable of removing dried mud as well.

The spare wheel is stowed separately beneath the rear luggage compartment,

Always renowned for its accessibility, the new engine retains all the desirable features of its predecessors.

xe four-door saloon

The Morris Minor is capable of carrying four adult passengers in comfort and this interior view demonstrates the well planned interior arrangements.

and the latter is capable of housing one very large suitcase, a sizeable dressing case and a large quantity of odds and ends. A further advantage is the boot lid which can be opened right up and laid against the roof of the car, thereby helping greatly the ease of access. In bad weather the lid can be propped in the more normal position for unloading purposes.

We have kept the description of the handling qualities of the Minor until the end because they are the real crux of the matter. The little car feels from the start like a thoroughbred. The rack and pinion steering is beautifully light and precise. The suspension is firm at speed and yet the ride is never harsh. The brakes are fully up to the new performance potential, and a deliberate attempt to induce fade on a three-mile descent proved fruitless. The close-ratio gearbox has a charm of its own, and seems to have inherited something of an M.G. ancestry. Fifty m.p.h. in third is well within the range when hurrying. It is refreshing to find a family saloon in which the shifting of the gears provides so real a pleasure.

Not all who purchase the new Morris Minor will make full use of its remarkable performance but those who do may be interested to know that driven flat out at dawn over some of the faster roads in Britain we achieved a petrol consumption of 34.8 m.p.g. at an average speed which would do credit to a sports car. In such conditions the noise emitted by the horn, which is reminiscent in its sound to that of a London taxi, caused us some anxiety lest lorries about to be overtaken failed to register. It should go on record that this rather curious note possesses remarkable powers of penetration as well as being sufficiently subdued for urban use and that, in consequence, we were never embarrassed by any inability to give adequate audible warning of our approach.

The Minor is a very ready starter even when left out all night and has an ability to warm up quickly which is most praiseworthy. Nevertheless, rather in keeping with the best traditions of the Nuffield range, the little engine does not develop anything like its full power for the first mile or two, after which a most satisfying output suddenly becomes apparent. Thereafter the power unit turns over throughout its range with a degree of smoothness and sweetness which is most commendable in any car and seldom encountered in one belonging to the lower price group.

The Minor has an excellent turning circle, and is in consequence a delight in traffic. Both mechanically and from the coachwork point of view everything has the appearance of being well and conscientiously made. The alligator bonnet reveals an engine which it is supremely easy to reach, and in fact the Minor 1000 may be summed up as the answer to those who need the economy and comfort of the traditional small saloon allied to a responsiveness and "gameness" which has always been the prerogative of the well-bred sports cars of the world.

Specification

Engine

Cylinders	4
Bore	62.9 mm.
Stroke	76.2 mm.
Cubic capacity	948 c.c.
Piston area	19.29 sq. in.
Valves	Overhead (pushrod)
Compression ratio	8.3/1
Carburetter	S.U.
Fuel pump	S.U. (electric)
Ignition timing control	Centrifugal and vacuum
Oil filter	Tecalemit full-flow
Max. power (net)	37 b.h.p.
at	4,750 r.p.m.
Piston speed at max. b.h.p.	2,375 ft./min

Transmission

Clutch	Borg and Beck s.d.p.
Top gear (s/m)	4.55
3rd gear (s/m)	6.415
2nd gear (s/m)	10.8
1st gear	16.47
Reverse	16.47
Propeller shaft	Hardy Spicer
Final drive	Hypoid
Top gear m.p.h. at 1,000 r.p.m.	15.18
Top gear m.p.h. at 1,000 ft./min. piston speed	30.5

Chassis

Brakes	Lockheed hydraulic
Brake drum internal diameter	7 in.
Friction lining area	63.8 sq. in.
Suspension:	
Front	Torsion bar, independent
Rear	Semi-elliptic
Shock absorbers	Armstrong hydraulic
Steering gear	Rack and pinion
Tyres	5.00—14

Coachwork and Equipment

Starting handle	Yes
Battery mounting	Bulkhead (beneath bonnet)
Jack	Smith's Steadylift
Jacking points	One (central) on each side
Standard tool kit	Starting handle, wheel brace, jack, grease gun, screwdriver, box spanner and tyre-pump.
Exterior lights	Two head, two side, two tail, two stop.
Number of electrical fuses	Two
Direction indicators	Semaphore type
Windscreen wipers	Electric self-parking
Sun vizors	Two
Instruments	Speedometer, fuel gauge
Warning lights	Ignition, oil pressure, main beam, direction indicators.
Locks: With ignition key	Off-side front door, luggage boot
Glove lockers	Two
Map pockets	None
Parcel shelves	Two (below instrument panel and behind seat)
Ashtrays	Two
Interior light	Yes
Interior heater	Yes
Car radio	None
Extras available	None
Upholstery material	Leather, with leather-cloth on non-wearing surfaces
Floor covering	Pile carpet
Exterior colours standardized	Seven
Alternative body styles	Two-door saloon, convertible, and station wagon

Maintenance

Sump	7 pints (incl. filter) S.A.E. 30 (summer), (20 winter)
Gearbox	2¼ pints, S.A.E. 30
Rear axle	1½ pints, S.A.E. 90 (80 for extreme cold)
Steering gear lubricant	S.A.E. 90 (80 for extreme cold)
Cooling system capacity	9¾ pints (2 drain taps)
Chassis lubrication	By grease gun every 1,000 miles to 10 points
Ignition timing	2 degrees B.T.D.C.
Spark plug type	Champion NA8, 14 mm.
Spark plug gap	0.024-0.026 in.
Valve timing	I.o., 5° B.T.D.C., i.c., 45° A.B.D.C.; e.o., 40° B.B.D.C.; e.c., 10° A.T.D.C.
Contact breaker gap	0.014-0.015 in.
Tappet clearances (cold):	
Inlet	0.012 in.
Exhaust	0.012 in.
Front wheel toe-in	$\frac{3}{32}$ in.
Camber angle	1 deg.
Castor angle	3 deg.
Steering swivel pin inclination	7½ deg.
Tyre pressures:	
Front	22 lb.
Rear	22-24 lb. according to load
Brake fluid	Lockheed
Battery type and capacity	Lucas 12-volt, 38 amp./hr.

ONE OF THE LOWEST PRICED OF THE BRITISH IMPORTS, THE MORRIS MINOR COMBINES EXCELLENT ECONOMY WITH GOOD HANDLIN

MORRIS MINOR TEST...

FEW CUBIC INCHES but with high compression enables the little engine to hold its own in typical U.S. high-speed traffic.

THE Morris-Minor has been around for quite a long time. It is one of the makes turned out by the British Motor Corporation (MG, Austin and Austin-Healey) and was first introduced into this country along with the MG. However, it never enjoyed a very big sale in the U.S., largely because it never was highly advertised nor pushed very hard by many dealers.

According to latest reports, this de-emphasis is about to change. Odds are car buyers will be hearing much more about the Morris-Minor 1000 in the future.

Engine changes throughout the years have given the Morris increasing performance so that it is now about at the point where anyone considering a super-economy car might very well give the little vehicle some attention.

A new 58 cubic inch (948 cc) overhead valve inline four makes the car fairly competitive in the economy car category. For instance, the current Morris w outperform both in top speed and, to lesser degree, in acceleration, the Germ Volkswagen.

The styling of the car, being typica British, is one of its greatest handica The shell is substantially the same o that has been used for quite some tim For those who like the tweedy flavor, may go over big.

The driver's position behind the wh is something like the erect posture sumed in a pickup truck. The steeri rim is low and at an angle, which is r at all bad and the forward visibility good.

The instruments, located in the cen of the dash, include needles for spe ometer and gas, lights for oil and ge erator. But nothing has been suppli for water temperature, which may not important in cool and foggy Englar but in many export areas is not or helpful but essential.

Elsewhere on the control panel are t pull lever for a starter, choke, li knobs and the wiper lever. The horn on the turning signal indicator—this dicator, incidentally, was not the se cancelling type, which is botheri Further, the turn indicator light is positioned for easy visibility, nor is clicker loud enough for good remindi qualities after a turn has been negotiat

The interior is British, meaning f leather, good chrome, but a dull-looki dash. There are two glove boxes of av age size, plus a huge package tray running beneath the dash which could carry additional miscellaneous items.

Seating in front is strictly for two, with the gear lever mounted on the floor in the middle. It has the customary four forward speeds, but the lever is a little on the short side for some short-armed drivers. If optional lengths were provided, it would not be a bad idea. Synchromesh is on second, third and fourth gears.

The rear seat is comfortable for two, although legroom is at a minimum. Trunk capacity is ample by small-car standards. The body is unit construction and is free of rattles.

The handling qualities of the Morris, considering its relatively high center of gravity, is especially good. The Morris has had torsion bars for years on the front, with semi-elliptics at the rear. The spring rate is a nice compromise between soft and firm riding. The brakes seem to be quite adequate for the car's weight.

When driven in normal city traffic, including expressways, there is no trouble keeping up with any pace, although the little engine admittedly is really turning over when cruising is at around 65 mph. Acceleration from 0-60 mph is 27.7 seconds.

The best features the Morris has to offer is good service facilities, since the car is nationally distributed through all MG dealers, along with its excellent gas mileage. The average for this test, including fast expressway driving and traffic, was 35.5 mpg. Under better conditions, 40 mpg would not be unusual.

Naturally, with its small displacement, it takes some gear box manipulation to stay with other cars on the road. But a good many drivers find this enjoyable and **feel that it puts some of the fun and/or sport back into driving.**

Those who like the foregoing qualities, along with British styling, will find that the car goes better than some of its rivals. Although for a little more money, even quicker performance can be had, but if price and economy are your dish, check with your Morris dealer. •

MORE ROOM with two glove boxes and a full-width package shelf beneath the dash is an unexpected feature in this small car.

MORRIS TEST DATA

Test Car: 1957 Morris-Minor 1000 DeLuxe two-door sedan
Base Price (West Coast): $1595
Engine: 57-cubic-inch four-cylinder in-line
Compression Ratio: 8.3-to-1
Horsepower: 37
Dimensions: Wheelbase 86 inches, width 61, height 60, widest tread 51
Curb weight: 1650 lbs.
Transmission: Conventional four-speed synchro-mesh in upper gears
Acceleration: 0-30 mph 6.8 seconds, 0-45 mph 14.5 and 0-60 mph 27.7
Top speed: 75 mph
Speedometer Corrections: indicated 30, 45 and 60 mph are actual 29, 42 and 56 respectively

ABOUT A DOLLAR PER POUND is what the Morris Minor costs in standard trim. Maneuverability is one of its strongest points, makes it exceptionally easy to get in and out of tight traffic and parking spots. Top speed is a rather surprising 75 miles per hour.

ON TOP of the Rocky Mountains, at the end of the second-gear climb to the 14,260 foot high summit of Mount Evans, Colorado.

52 DAYS with a MORRI

11,500 Miles of Motoring Through 27 States of the U.S.A.

Recorded by JOSEPH LOWREY, B.Sc. (Eng.)

THANKS to the educational activities of Mr. Noël Coward, the strange habits of Mad Dogs and Englishmen are by no means unknown to the American public. To the average American, the idea of going all the way across his vast continent by car for the fun of it, when an airliner will fly you from New York to California for $99 while you try to sleep, is a little bit eccentric, the sort of silly trip that some people make once in a lifetime. The idea of travelling any great distance in a car of European size still seems to most Americans to be even more unintelligent than driving across the continent in a real car. And as for going from New York to California in a Morris Minor, only to return again in the same car by the longest and least direct route possible—just plain nuts, we gathered, was the near-unanimous American verdict on folk who did that sort of thing.

BEATEN by 48 hours, the Minor passed through Moscow (Michigan) going West two days after *The Motor* correspondents B & K had passed through Moscow (Russia) going East.

As a matter of plain fact, I never did intend to make the double crossing of the North American Continent in a Morris Minor 1000. Looking for a car with long legs and a short thirst for fuel, I decided to take advantage of my opportunities as a journalist to borrow a Wolseley 1500. But, the Wolseley 1500 proved to be one of the few good things of this world which could be bought with pounds but not as yet for dollars; so instead it happened that on the morning of July 13 I drove out of a rain-swept New York in a Morris 1000 two-door saloon of which the speedometer read 00742 miles. When I reappeared in New York around 1 a.m. on September 3, even the folk at Hambro Automotive Corporation who import Morris cars into the U.S.A., seemed faintly surprised to find a speedometer reading of 12,244 miles on their car—and when it was mentioned to them that a British immigrant who was arriving in New York the next week had been advised by my fellow-traveller Edgar Wadsworth to explore the possibility of buying this same car, as a nicely run-in model in which to depart towards California with his wife, Wojtek Kolaczkowski who looks after the Hambro showroom on 57th Street, gave us a look suggesting that he thought maybe Noël Coward was right about Englishmen.

Mad? If I get the chance to repeat the trip next year I shall not need asking twice; though just possibly I may try to choose a time of year when the mercury in the thermometer rushes past the 100° F. mark a little less frequently. From the red-tinted backs of our white nylon shirts, it would seem that some of the weather we drove through was hot enough to make even a dead cow perspire, refrigerated interiors being one thing which we

INOR 1000

DOWN in the hot valley at Zion National Park, the Minor carries an evaporatively-cooled canvas drinking water bag in case of a puncture delay on a lonely desert road.

envied some of the locals. For the rest, we seemed to cover at least as many miles as most other people between 9.30 a.m. and supper time, and to arrive at least as fresh and hungry as anyone else—and arriving by Minor 1000, we had a whale of a lot more dollars left for supper than the V-8 voyagers who were not all getting 10 miles out of their little American gallons of gasoline.

Statistically, our 11,502 miles (on a distance recorder accurate to two car-lengths in 30 miles of Turnpike) cost us $113.45 for 331.3 U.S. gallons of regular-grade petrol (starting and finishing with a full tank), an overall average of 34.7 miles per U.S. gallon equivalent to 41.7 miles per British gallon; this was a modest 0.985 cents per mile or 0.845 British pennies per mile for fuel which cost us an average of 34.3 cents per gallon—much of our fuel was bought in remote mountain areas where it sometimes cost as much as 40 cents per gallon, just double the rock-bottom price for which we got one tankful in Missouri. Only once, in California, did we spend money on premium-grade petrol, the less heavily leaded cheap grades proving entirely adequate in octane rating.

Oil cost us $8.43 for 18 American quarts to top-up the sump, plus four changes of oil which used 3 American gallons of oil costing $6.06, together adding just under 13% to our fuel costs. Greasing at roadside garages at rather more than recommended 1,000-mile intervals cost $8, or about 7% of our petrol cost, and repairs cost us just 10 cents (eightpence halfpenny in English money) for one new tyre valve, 0.097% of our fuel cost. On the whole, we think we travelled fairly cheaply at 1.18 cents per mile (1.015 pence per mile) total motoring cost, and occasional Turnpike roads at a toll of 1¼ cents per mile looked jolly expensive luxuries to us—we, in fact, paid out $11.55 in toll for using various bridges, tunnels and turnpikes plus $8.50 toll to take the car through Grand Canyon, Zion, Yosemite, Craters of the Moon and Yellowstone National Parks.

Will the smallest Morris cope with American conditions? Our 10 cent repair bill in 11,500 miles hints at the answer to that question. In addition to this replacement, I reset the contact breaker gap after 4,500 miles, this having closed up and made the performance sluggish by the time we reached Los Angeles. I also tightened the oil filter and oil pressure relief valve cover nuts to check slight oil leakage, adjusted the brakes after 6,000 miles to restore them after initial bedding-down, and used a Phillips screwdriver (not included in the tool kit, surprisingly) on a sun vizor bracket and the door catches to check slight rattles. Both wires had to be coupled to the stop-lamp switch when I discovered that our stoplamps were not working—I doubt if the leads had ever been coupled up previously—and at intervals I

SUNSHINE was occasionally interrupted by thunderstorms, one of which flooded the main road through Otis, Colorado.

SCENIC wonder of Arizona, the Grand Canyon of the Colorado River was shrouded in heat haze.

HISTORIC relic of a w
coach are preserved at
below the long

52 DAYS with a MORRIS MINOR 1000

picked up the handbrake release knob from the floor and screwed it back into place, none of these quick and simple jobs involving paid help. The general view of garages which greased the Morris and could find nothing else to do to it seemed to be, that if many customers changed over from complicated Detroit automobiles to straight-forward imported models the repair trade would catch a cold. . . .

Will the Minor stand heat? I think that our hottest weather was about 108° F. in the Nevada deserts around Las Vegas, and despite a butterfly-plastered radiator we had no overheating, the electrical petrol pump ticking rather as if it was trying to keep Double Summer Time on certain occasions, but the engine always starting first touch whether cold or hot and never faltering. At one stage engine oil consumption rose to around 1,000 miles per U.S. gallon (1,200 per imperial gallon) with mixed brands of S.A.E. 30 oil in the sump, causing us to change over to S.A.E. 40 oil and avoid mixing brands, after which our last 3,000 miles of driving called for only one American quart of oil each 1,000 miles.

How does a 37 b.h.p. car cope with a morning rush-hour in which every other car has upwards of 100 b.h.p. on tap? Ten days spent 20 miles out of Detroit, making daily visits to factories in and around the city showed that, with reasonably energetic use of the gear lever but not over-revving or exceeding speed limits, the Minor got through the morning and evening rush hours quite as fast as most other folk, losing a car's length when the light first went green and the Yank alongside span its rear wheels, but usually hitting the local cruising gait of 50 m.p.h. without dropping back appreciably further. Of the European cars I saw in city traffic, a large proportion were darting from lane to lane in a manner which American cops are apt to frown upon but obviously were penetrating the traffic just as nimbly as they do in the less tidy congestion of European towns.

Will the Minor tackle big mountains? Our high spot was the summit of Mount Evans, Colorado, which at 14,260 feet claims to be the highest motor road in the Northern Hemisphere and tops the highest Alpine pass by some 5,000 feet, three of us riding to the summit non-stop without anything lower than 2nd gear being needed—we had already left our luggage at our Motel in Idaho Springs on that occasion. Even here, as on other occasions at 10,000 and 12,000 feet altitudes across the Continental Divide, the engine would still idle in a slightly ragged but quite reliable fashion, without any re-setting of the mixture. On long, straight grades up into the mountains, there were infuriating occasions when 37 b.h.p. less some loss from altitude sufficed only for perhaps 45 m.p.h. in top gear, or the pace might even sink to 35 m.p.h. in 3rd gear, what time big American cars sailed by—the occasions were infuriating because in at least 75% of the cases the cars which passed us would have to be re-passed on winding or downhill road ahead, and once re-passed might never be seen again. The "1000" engine gets across mountains

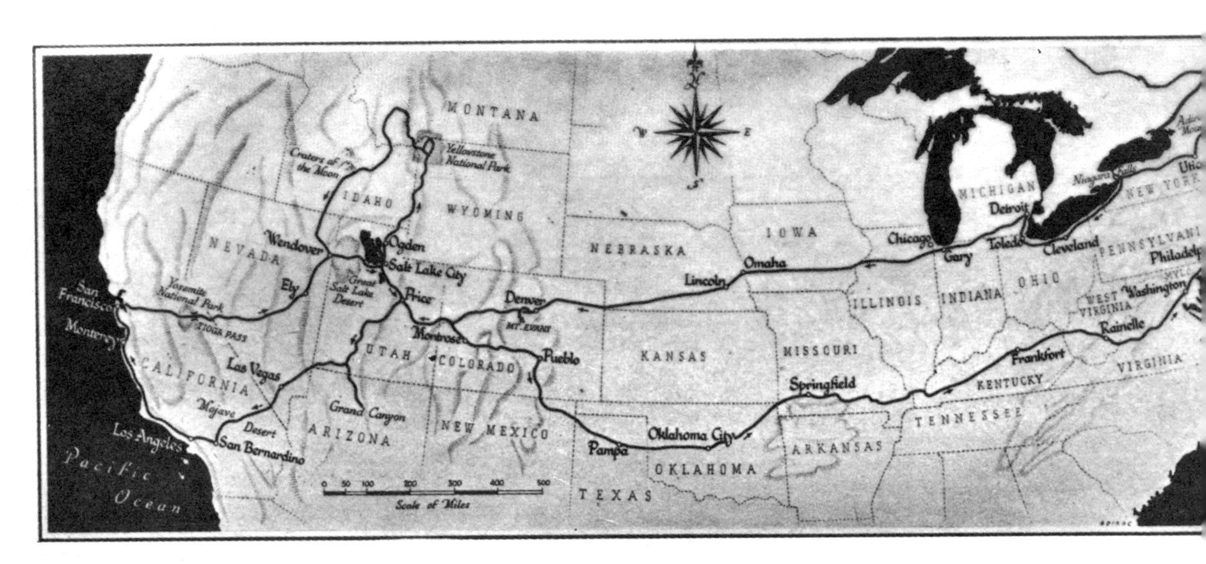

UNIQUE product of freakish nature, the salt flats outside Wendover, Utah, look vast and lonely even when speed record attempts are in progress.

ɔmotive and a passenger d depot in Idaho Springs, ; of Mount Evans.

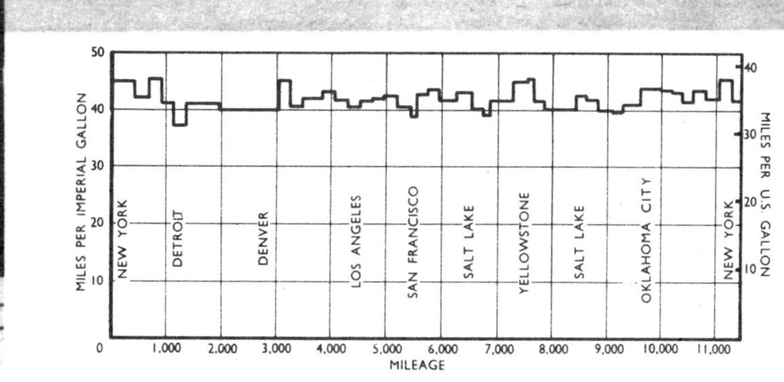

quite nicely, but a "1500" engine in about the same size of car would have had advantages, both in the mountains when altitude and gradient were adverse, and sometimes too in the plains when an adverse wind could blow all day and make our 60-65 m.p.h. cruising pace into almost a full-throttle maximum.

But, cruising at just over 60 m.p.h. most of the time, we found the little Morris well able to cover big distances. Our longest day runs were 586 miles from Montrose, Colorado to Pampa, Texas and 536 miles from Rainelle, West Virginia to New York, N.Y., both runs from late starts and including not merely ordinary meals but in each case loss of an hour due to Eastward driving across time zone boundaries. Our best accurately noted average speed away from the Turnpikes was 120 miles in a few seconds under 120 minutes without exceeding 65 m.p.h., between Ely, Nevada and Wendover, Utah.

In Texas, our mile-a-minute-plus cruising gait was below the local average, but there were other areas such as Kentucky where we seemed to be the fastest car on the road—driving habits vary widely from State to State and County to County. In hot weather, the fact that the Minor is inclined to be draughty internally is no disadvantage, and with windows wide open the growl of a rather noisy rear axle was not conspicuous as it might be in winter. The individual front seats proved very comfortable, and when the third member of the party stayed behind to take a job in Los Angeles we improved the limited range of driving seat adjustment (and the lack of passenger seat adjustment) by moving both seats to the more-rearward alternative mounting positions which are concealed beneath the carpets of every Minor.

In New York before I started on this trip, a local motoring writer described American roads as rough. On the whole I disagree, for whilst there are badly surfaced dirt roads in many residential areas, even the least important through-roads passed fairly smoothly beneath our not-unduly-soft torsion bar front springs. Our roughest road was a dirt one through the fringes of an atomic testing area which we used as a short-cut from Yellowstone to the Craters of the Moon volcanic area, and our most strenuous route the Tioga pass in Yosemite Park where 9,900-foot altitude and gradients steeper than 1 in 7 would have made a stop-and-restart by the luggage-laden Minor very doubtful in a few places. Somewhat to my surprise and much to my joy, our trip hardly reduced shock absorber efficiency at all.

Around American cities, the compact size of the Minor did not prove such a great advantage as the clumsiness of big cars in European cities had made me expect. At times a narrow car could filter round the corner at traffic signals when there was

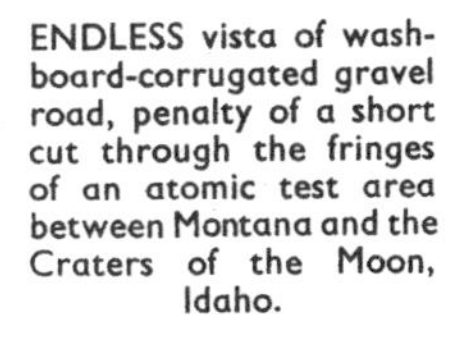

ENDLESS vista of washboard-corrugated gravel road, penalty of a short cut through the fringes of an atomic test area between Montana and the Craters of the Moon, Idaho.

52 DAYS with a MORRIS MINOR 1000

WESTERN landmark of the journey, Golden Gate bridge spans an entrance to San Francisco harbour which is strangely reminiscent of Falmouth Bay, in Cornwall.

not room for anyone else to follow, and occasionally a short car would fit into an otherwise useless spot in an apparently full parking lot. But in the main, American cities are designed for big cars, and with kerbside parking limited to defined areas with one car per parking meter the space not occupied due to the smallness of the Morris could not be used by anyone else.

It was on the open road, oddly enough, that at times we were happier in a small car than we would have been in a large one, for in many areas (Missouri seemed a typical one) quite busy roads and bridges seemed uncomfortably narrow for big cars meeting one another at speed. Crossing most of the continent from Detroit to Los Angeles with three people in the car, packing all the luggage in was a job which needed to be done tidily, but for two people the Minor was big enough to permit some change of position during long drives. At first, driving in shirt sleeves, we missed the coat hooks which are to be found above the windows of every American car, but we soon found out that the Minor's steel body structure had a lip above the windows, from which coat hangers would hang securely with our jackets upon them.

Because the rear suspension has been stiffened or for some other reason, the present-day Minor does not seem to handle with quite the phenomenal precision of earlier and slower examples with side-valve engines; this I am sure is fact, and not merely an impression resulting from everything else having improved so that standards of judgment are higher. But the rack and pinion steering remains quick and positive, and by American standards the Minor handles like a sports car, rolling a little when cornered fast, but on open curves habitually overtaking in virtual silence outside large vehicles from which the most anguished shrieks of rubber in torture were emerging. Usually we performed this manoeuvre with the driver's spare arm dangling out of the window in the not-so-cool breeze, just to rub salt in the wounds of the overtaken driver who was busy winding up yards and yards of steering with every available hand. On a cambered or windy straight, however, the Minor actually needs a firmer one-handed hold of the wheel than does many a modern American car, driving conditions in which cars come out of Detroit 6-abreast at a steady 50 m.p.h. during the evening rush hour, on a road divided by painted lines into six none-too-wide lanes, having forced the evolution of American cars which really will follow a straight course without much regard for camber or the suddenly-encountered slipstream of a 55-60 m.p.h. Greyhound motor bus.

SCRATCHES on the Minor's paintwork were not all incurred in car parks, some also resulting from this Yellowstone Park bear's anxiety to get at the tin of salted peanuts!

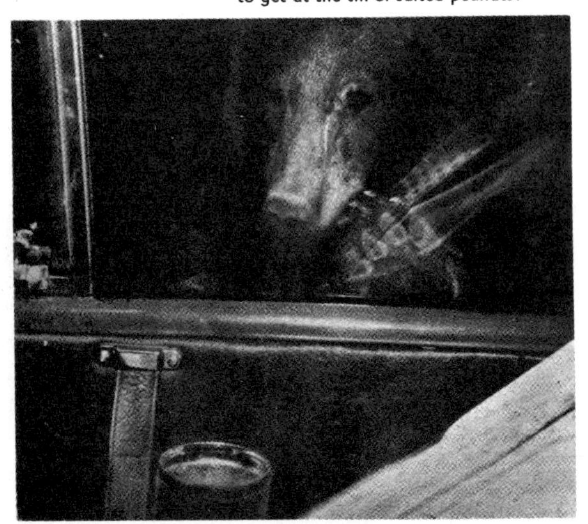

By mixing business with vacation in a manner which, apart from the irresistible temptation to add more and more visits into the programme, was highly successful, I was enabled to sample the Minor in very varied parts of America. On the first stage from New York to Detroit, a week-end got lost in the Adirondack Mountains and on the Canadian border at Niagara. Between Detroit and Los Angeles, we took in the highest mountain road and the deepest mountain canyon in the U.S.A. as well as the Mojave Desert, carrying on the front of the car the fashionable canvas "Desert Bag" of evaporatively-cooled water in case a puncture should delay us amidst the lonely, sun-scorched wastes which still await irrigation. Our route to the Salt Flats of Utah took in a Sunday run across San Francisco's Golden Gate Bridge as well as more mountains. Wet weather on the salt flats gave us time off to gape at the geysers and boggle at the too-tame bears of Yellowstone. The National Hot Rod Association's Drag-Racing Championships diverted us south and tempted us to see the unhurried life of Kentucky and Virginia during our final week-end. Mexico and the Alaska highway had to be left out of the schedule, but if anyone thinks that perhaps these would disclose limitations upon the Minor's performance, then two volunteers to go and disprove the heresy would be easy to find.

The object of this article is not to tell readers how to make a holidaymaker's allowance of £100 cover a dollar holiday; that story must wait until the Motor Show is over. Suffice it for now that a good British small car can laugh at American road conditions; that if you can spare three weeks for holiday-making in America plus about a week each way for the transatlantic boat trip, and are prepared to sleep two to a room in comfortable but out-of-the-cities Motels, you can apply for shipping accommodation in the sure knowledge that the personal and car currency allowances of £100 per person and £35 per car will not leave you too poverty-stricken a tourist to enjoy yourself in America just as much as we did.

STATISTICS

States visited: Arizona, California, Colorado, District of Columbia, Delaware, Idaho, Illinois, Indiana, Iowa, Kentucky, Maryland, Michigan, Missouri, Montana, Nebraska, Nevada, New Jersey, New Mexico, New York, Ohio, Oklahoma, Pennsylvania, Texas, Utah, Virginia, West Virginia, Wyoming.

Distance covered: 11,502 miles.

Fuel used: Regular-grade, approx. 90 Octane Research Method Rating, 331.3 U.S. gallons at average price 34.3 cents per gallon. (Equivalent to 276.1 British imperial gallons at 2s. 11½d. per gallon.) Average consumption, 34.7 miles per U.S. gallon; equivalent to 41.7 miles per British imperial gallon.

Oil used: (S.A.E. 30 and S.A.E. 40.) Topping up, 4½ U.S. gallons; equivalent to 3¾ British imperial gallons. Re-filling after oil changes approx. every 3,000 miles; 3 U.S. gallons equivalent to 2½ British imperial gallons.

Costs:	$	£	s.	d.
Fuel	113.45 =	40	10	4
Oil (including changes)	14.49 =	5	3	6
Routine greasing	8.00 =	2	17	2
Repairs (new tyre valve)	.10 =			8½
Total	136.04 =	48	11	8½

Fuel cost, 0.985 cents/mile (0.845d. per mile)
Total cost, 1.180 cents/mile (1.015d. per mile)

ROAD IMPRESSIONS

THE LITTLE MORE

—and how much it is

FUN WITH A SUPERCHARGED MORRIS MINOR 1000

IN racing circles, supercharging as a means of increasing performance is less popular now than it used to be; of course, concentration on the development of normally aspirated engines for racing—since the war, Grand Prix formulæ have favoured the unblown engine—has a great deal to do with it. But such matters need not concern the owner of a family car who wants a little more performance. Nor does he need to consider the high boost pressures of racing engines, which according to the type of supercharger employed, may reach 30 lb sq in. Pressures of 4 to 8 lb sq in above atmospheric give a considerable increase in power by reason of the improved filling of the cylinders. Modest forced feeding can bring considerable returns from a normal production engine without detriment.

It is difficult to generalize on effects of supercharging on reliability and rate of wear. Obviously it would be folly to supercharge a well-worn or unroadworthy model, while some designs are more suited to, and give greater gains from, the treatment than others.

Low-pressure supercharging, although increasing maximum cylinder pressures, gives increased power output at lower engine speeds compared with an unsupercharged engine with raised compression ratio. Increased wear and tear, when it does take place, can often be attributed to the driver consistently taking advantage of the enhanced performance, and driving the car harder and faster than is usual.

Greater heat flow to the water jackets is inevitable, but most cooling systems are well able to take care of this, and overheating is rare. In many instances, improving the breathing without resorting to supercharging results in better torque at high r.p.m., though at the expense of output at lower speeds. With supercharging this disadvantage is not experienced; in fact, flexibility and low-speed pulling are invariably improved. Mixture distribution inequalities between one cylinder and another are overcome by the supercharger, which also aids the mixing of fuel and air and gives a higher rate of gas flow in the induction manifold.

Higher final drive gears are often desirable if the best use is to be made of the extra torque. The supercharged Morris Minor 1000 which is the subject of this article has been fitted with an alternative 4.2 to 1 crown wheel and pinion in place of the standard 4.55 to 1 unit. With its excellent steering and stability, and delightful, central-lever gear box, the Minor is rewarding material to work on. The owner, D. Griffiths-Hughes, a precision engineer who lives in Carshalton, is a believer in supercharging. He has also a DB2-4 Aston Martin, to which he has fitted a blower—a very potent combination which will be described on another occasion.

Work carried out on this Morris engine has not been confined entirely to the fitting of the supercharger. After the engine had been run in, an Alexander-modified cylinder head was fitted; the normal work of reshaping and polishing combustion chambers and ports was done, but the compression ratio was left at 8.3 to 1. A plain copper gasket was fitted. The 4.2 to 1 final drive was also obtained from Alexander Engineering.

Modification of the exhaust system consisted of fitting a Servais three-branch manifold, and the hot-spot was omitted —a straight outlet being arranged from the siamesed ports for cylinders two and three. A larger than normal tail pipe was supplied, the silencer being shortened by 6in and fitted at the extreme end. This arrangement was found to give an increase in tractive effort of 20 lb per ton, by Tapley meter, compared with mounting the silencer in the normal position.

Owner D. Griffiths-Hughes looks on while his blown Minor sets off for performance testing, with the Editor at the wheel

The supercharger selected was a Roots type manufactured by the North Downs Engineering Co., who supply superchargers for many makes of cars. They had not previously installed one in a 1000, though they had done so for the earlier model. This meant that a little adaptation was needed, mainly to accommodate the larger carburettor of the latest engine, the original instrument being used. The supercharger, manifold, brackets and oil pipes were supplied at a cost of about £80.

Drive to the blower is by rubber vee-belt from the front of the crankshaft, with a free pulley on the slack side to adjust

Served by the original S.U. carburettor, the Marshall-Nordec supercharger fits snugly on the left side of the engine. Below it is the Servais three-branch exhaust manifold

On the right of the blower can be seen the vee-belt drive with tensioning pulley and oil feed pipe. The blow-off valve is on the left. Accessibility of plugs, distributor and dip-stick is unimpaired

tension. A copper pipe is taken from the oil pressure gauge union on the cylinder block to a restrictor on the casing, for metered lubrication of the drive gears. A drain pipe allows surplus oil to return to the sump. The boost gauge was an extra, costing £2 5s. Driven at 1.5 times engine speed, the supercharger gives a maximum boost of 8 lb sq in. The standard breather in the rocker box cover is connected by flexible pipe to the intake side of the carburettor. It was not found necessary to have special plugs.

Starting from cold seemed just as easy as on the standard Minor under similar weather conditions. Tick-over when warm is at a slightly higher speed than usual; with most supercharged cars it is not possible to make them idle very slowly.

It was soon obvious, from the beginning of a run up A1 to Huntingdonshire, that acceleration was greatly improved. High revs were not needed to get results, though it was noticed that near the maxima in the gears the engine smoothness evident at lower speeds was maintained.

There was audible evidence of the supercharger gears at work —a pleasant whine most noticed during acceleration, followed by a quieter air intake roar as speed built up. These signs of the machinery at work were not loud enough to be tedious on a long journey and cease to be noticed at all after an hour or two on the road. It was difficult to hear the car radio, but this was mainly because of body resonance from the tyres on certain surfaces, and wind noise. The Editor said that it all reminded him of the sounds inside the cockpit of a Meteor Fighter when the engines are being run-up.

Cruising speed was an effortless and comfortable 60-70 m.p.h., aided by the higher-geared back axle. Delightfully responsive to the throttle, even if slammed wide open, the engine could not be made to pink on premium fuel.

It was instructive to watch the antics of the boost gauge—flicking from full minus boost on overrun to a positive reading during acceleration, depending on the throttle opening and engine speed at the time. Zero boost was seen at about 50 m.p.h. in top gear and the highest reading obtained was plus 8 lb sq in during acceleration tests.

The engine tended to run too cool even during hard driving. With part of the radiator blanked off, the highest reading shown on the thermometer was 165 deg Fahrenheit.

THE LITTLE MORE . . .

Other motorists, unaware of the Minor's hidden talent until either left behind as the traffic lights changed or passed unexpectedly, often showed their surprise, some being spurred on to greater efforts themselves. Drivers of heavier metal sometimes looked quite offended.

A disadvantage of cars that go faster than it appears they should is that approaching or waiting drivers may misjudge the closing rate, and due allowance has to be made for this.

A direct performance comparison with the standard Minor 1000 is given in the accompanying table. It should be remembered that the blown car is higher geared than the standard one and has plenty of r.p.m. in hand at its maximum level speed. An improvement of 12.2sec in the standing start acceleration from 0-60 m.p.h., and a reduction of 2.8sec in the elapsed time for the standing quarter mile, give a good idea of the potency of the supercharged car.

That improvement is not confined to performances when maximum revs are used is shown by the shorter times for accelerating from 30-50 m.p.h. and 40-60 m.p.h. in top gear, for example. This is evidence again of the excellent flexibility and low-speed pulling power of the supercharged engine. It was clear that there would be some improvement in maximum speed, but conditions were against getting the best results. However, during a run of limited distance with slight assistance from the wind, a true 80 m.p.h. was obtained, and 76 m.p.h. in the opposite direction, a mean of 78 m.p.h. This was when carrying two occupants and test equipment. These top-speed figures could certainly be improved upon, given a longer clear stretch in which to work up speed.

Chassis-wise, the Minor does not seem to be at all over-taxed by this greater performance. Brakes are well up to their task, and road-holding—a strong point with the model—gives confidence. It is always pleasant to enjoy again the Minor's steering—precise and accurate, though the tremor common to rack and pinion steering shows itself at higher speeds, indicating the importance of correct wheel balance.

In towns one can trickle through at low engine speeds in third or top gear without attracting any attention, and generally the car is so tractable, quiet and free from fuss that shopping and similar excursions are all part of its daily work.

Fuel consumption is affected by supercharging. The figure we obtained over a distance of about 138 miles of mainly hard driving was 26 m.p.g. With the standard car, under similar road conditions, 36 m.p.g. would probably be obtained, though not with such speed or acceleration, of course. The blown car will in turn approach 30 m.p.g. if driven less hard.

Mr. Griffiths-Hughes believes in using a mixture slightly richer than normal on his supercharged cars—a precaution to safeguard pistons and valves—and his air-fuel ratio of 12 to 1 was determined with the use of an exhaust gas analyser. He produced the correct profile on the S.U. carburettor needle himself. Certainly the car has given much pleasure to him and also to his wife, who uses it daily. We, too, were glad of the opportunity for a day's motoring in it, and sorry to relinquish it for less deceptive methods of transport.

D. M. P.

ACCELERATION

	Supercharged (4.2 to 1 axle)	Standard (4.55 to 1 axle)
From rest through gears to:		
M.P.H.	sec	sec
30	5.9	6.8
40	8.6	—
50	13.6	18.8
60	19.1	31.3
70	27.9	—
Standing quarter mile	21.4	24.2
From constant speeds:		
20-40 m.p.h. in **2nd**	5.1	—
20-40 m.p.h. in **3rd**	7.6	10.5
30-50 m.p.h. in **3rd**	8.6	11.7
30-50 m.p.h. in **top**	13.0	18.2
40-60 m.p.h. in **3rd**	12.3	—
40-60 m.p.h. in **top**	13.8	23.9
50-70 m.p.h. in **3rd**	18.9	—
50-70 m.p.h. in **top**	15.3	—

YOU DRIVE IT!

Small-car owner Tom Carroll, an accountant of Woodside, L. I., N. Y. tried out and gives you his report on the

ENGLISH MORRIS MINOR 1000

MUCH has been said in recent years, and especially in recent months, about the economy of small cars—and there is no doubt that these new imports stretch your dollar beyond expectations. The question, however, is how much of a sacrifice must be made to acquire this economy?

In my experience, no sacrifice is necessary. This is based on driving or living with two successive "small jobs" for awhile (two Austin A40's — *Editor*).

Since 1948 I have logged (he really logs 'em statistically too! — *Editor*) over 100,000 miles in extensive trips throughout the U.S. and Canada. I have found that these little cars furnish comfort as well as economy on these long stretches between eastern and western cities. Despite their limited horsepower, they are definitely turnpike cars.

Quite a while ago, I was approached by a man from SMALL CARS Magazine for a sort of carside interview on what I thought about small cars. We had a very lengthy conversation and I was asked if I was interested in any make other than Austin. I mentioned that I'd heard and read a lot about the Morris 1000 and said that sometime I'd look up a dealer and take a spin in one. My own Austin got badly banged up while parked and I'm waiting for an insurance adjustment.

A few days later I got a call from the magazine asking if I'd like to try out a Morris on a sort of road test for a couple of days. Unfortunately it was during the week, so my testing was confined to after-work driving, in what turned out to be very rainy weather.

Before picking up the Morris I was frankly sceptical that I could get the same performance from a small 1000 cc car (58 cubic-inch engine — *Editor*) as I could from my 1300 cc job (78 cubic inches). Time and mileage certainly proved otherwise.

To start with, the Morris 1000 has all the earmarks of a sports car. Its willing engine, in all gear ranges, gives the impression that the car was built for the sports-car enthusiast. The gearbox was a pleasure to use, as it allowed very quick shifts. Third speed gave nice zip in traffic and it could also be used to advantage for passing slow cars on the road. These were pleasantly surprising characteristics to find in a four-seater family sedan.

Not being a great believer in the accuracy of speedometer readings, and all of a sudden being a test driver, I decided to test-run the Minor between New York City and Spring Valley, by way of the N.Y. State Thruway. I have frequently made this run with various cars larger than the Minor and know the time the trip takes.

I felt sure that running the Minor over the same course would furnish me with the information I was seeking. It was hard to believe but, keeping within the speed limits indicated by road signs, I covered the distance in 10 minutes under my fastest previous run! I think that, out of sheer "shifting pleasure," I might have been using the gearbox more than I normally do.

This inspired further tests to find out what this little bomb could do on acceleration. It is hard to believe that

(*Continued on Page* 14)

Minor stationwagon looks good in any surroundings; large window area at rear has sliding panes. V. & L. have also perfected a stationwagon based on the '1000" model, which will be available before long.

ROAD IMPRESSIONS

More Versatility

LORD NUFFIELD S tiny Minor is a happy little soul with winsome ways, sweet manners, and a heart as big as the bucket that you'd use to wash it in. So what more logical a basis could there be for a pint-sized stationwagon (if you like stationwagons, and most Australians do) than that same ubiquitous little Morris Minor?

The trouble is, of course, that since Mr. Morris gave Junior a big "1000" engine to play with, there has only, until recently, been the one model Minor available in this country — to wit, the saloon car, in both 2-door and 4-door versions. And that hasn't suited the stationwagon builder's book at all, because the saloon Minor, you will recall, is a chassisless beast —which means his having to make all kinds of strutting, welding, and supporting operations before he finishes up with a platform strong enough to support the fancy new body's extra weight.

The "1000's"' older brother, on the other hand—though having a deal less shove at the back wheels, on account of a smaller engine and wider-spaced gear ratios — *was* admirably suited to such a conversion, because it was available in Utility form, with not only a good rugged steel-rail chassis, but also with a lower geared back end! And there lay the clue to the whole thing.

Well, Eric Lane, of Vaughan and Lane, Nuffield Distributors, was getting pestered by a lot of people wanting stationwagons and not being able to buy them under the Morris nameplate; and he put two and two together, and came up with the interesting little job depicted on these pages.

Eric simply latched on to a goodly supply of 8 h.p. Morris Minor Utes, flung away their tray bodies, and whopped them into the bodybuilding shop to produce the new body-beautiful. The resultant 'wagon is being sold by him through his substantial chain of Nuffield dealers at an all-in figure of £995—and there appears to be no lack whatever of takers.

Eric told us, too, that a lot of folks are now yelling at him loud and long for "1000" stationwagons, a fact that's been causing him to burn the midnight oil more than somewhat; and at time of writing he asked us to point out that he thinks he has at last overcome the problem of putting a body on a "chassis" where there ain t no chassis at all—so that any day now, his well-tested version of a Morris "1000" stationwagon should be available. (He expects it will cost best part of a hundred pounds more, though.)

(Since the above was written, the "1000" utility chassis has become available.—Ed.)

Be that as it may, Vaughan and Lane offered us one of their 8 h.p. Series II jobs for a few days, to see what we thought of it, and this article is the result. We picked the car up in the evening and pointed it towards our home, and the first thing we noticed was the fact that the little creature seemed to be revving its head off. (We're rather used to the high-geared "1000" model, you see).

Low Geared . . .

As the miles went by, however, we came to appreciate the car's low

Stationwagon carries four easily; six at a pinch. Spare wheel compartment and ditto for tools lie beneath flat floor; are readily accessible with vehicle laden.

Showing how tailgates are supported, and how a tall man can load the 'wagon easily for business or pleasure.

. . . *from* **REUBEN HALL**

for MINOR

A Sydney firm has produced this excellent stationwagon conversion for Morris Minor — at a cost of less than £1000 including tax. We tried it, liked it, can state that it will prove popular in this "wagon-minded land".

gearing more and more, for the acceleration and general upsurge through the gears, and from quite low speeds, was virtually a new experience for us, in a Morris Minor. Several times we diced playfully with Series II saloons, and even with "1000's", on the road, and we were gratified to be able to out-accelerate all of them both low down in the gears and in top gear!

Every dog has its fleas, however—and it is disconcerting to be restricted to a 40/45 m.p.h. cruising speed whilst other Morrises are loping past you at 50/60! This restriction is necessary because, at higher speeds, the little engine, through its low road gearing, is revving unduly fast; so fast that to cruise at 50 m.p.h. or beyond is tiring to both driver and motor. But you do feel the benefit as you romp up hills in top gear, that have the other cars lugging heavily or else grinding their way along in third.

And again, you pay for this mountain goat activity in the form of a slightly higher fuel consumption—34.3 m.p.g. was the figure we recorded over some 250 miles of fairly vigorous driving. *Continued*

This view shows how "ute" type tailgate has been incorporated at rear; good finish of the little vehicle's exterior woodwork.

MORE VERSATILITY FOR MINOR

Continued

Four Seater . . .

Taking first things first, a description of the conversion is in order; and it is primarily a 4 seater stationwagon body having two doors plus top and bottom tailgates. Main framework is in steel, with wooden garnish affixed over the rear cornerposts, waistline mouldings, the frame of the top tailgate, and around both rear mudguards. All side panels are in steel, and the lower tailgate is a neat adaptation of the strong, serviceable tailboard which normally forms part of the utility's body structure.

There is room for three small people abreast on the front seat, and for ditto at the rear—but a more comfortable journey would be had by all if the load were restricted to four adults or to an equivalent number of children.

Legroom, though not liberal, is adequate.

Can Sleep . . .

The tip-down rear seat is secured in its "up" position by two latches, and in the "down" position there is achieved a flat rear floor approximately 5 ft. 3 in. long. Full sleeping length can be obtained either by removing the front seat or by lowering the rear tailgate and building its height up to floor level. The rear floor is covered in durable rubber flooring material, and beneath it, and directly accessible from the rear, are two compartments—one for tools, and the other for the spare wheel.

Interior trim is in plastic leatherette, varnished wood, and (in the case of the head lining) washable plastic material.

There is a centrally placed interior light, a glove box on either side of the facia—each has its own lid—and lockable sliding windows alongside each rear seat. A pleasing ventilation can be obtained on hot days by driving with both front ventipanes opened forward and with the top rear tailgate held up and supported on its strut, and allowing the wind to rush through unobstructed.

Our three complaints concerning the stationwagon's interior are as follows: (1) The front seat is rather high, causing the head of a tall driver to touch the headlining when negotiating bumps, and rendering it difficult for him to discern traffic lights set at a high level; (2) Clutch, brake, and accelerator are set extremely closely together, so that a bigfooted driver can easily stomp on the loud pedal instead of the brake. This could be awkward at crucial times—although we admit it is a delightfully convenient placement for toe 'n' heeling. We suggest that V and L bend the accelerator arm slightly to the right, for safety's sake; (3) The difficult access (for women) to the rear seat. We feel that a little careful attention to design could improve this relatively easily.

Improved Handling . . .

On the road the little 'wagon endeared itself to us very quickly. All the old Minor handling characteristics are still there in full, and in fact they are of a slightly improved order due to the 'wagon's changed distribution of weight. Weight distribution is now about 48/52 per cent. (carrying driver only) in favour of the rear wheels; and it was found, when cornering with this loading, that steering was as near as dammit to neutral. With two persons in the front seat there appears the merest trace of oversteer—just sufficient to let you slip the car around a corner speedily on the edge of a tail-first slide; whilst, when carrying rear seat passengers, the oversteer is almost, but not quite, as pronounced as that of a Volkswagon.

The gearshift is positive enough, but the synchromesh can be beaten between third and top gears if the change is snatched too quickly. And my—it is only after driving this earlier type of Morris once again, with its long, central-floor shift, that one really appreciates the cosy little stub-shift of the "1000"! For, although this gear change is good—really good—it just isn't in the new model's class as a foolproof unit and that's all about it.

Over rough, corrugated roads the little 'wagon behaved herself admirably and had less axle hop than has the current saloon version; possibly due again to the greater amount of weight lying towards the rear. The steering was quick and precise always, though still with that trace of rack-and-pinion "flutter" that is invariably associated with small Morrises.

Body Roll . . .

Body roll on corners was, to our mind, excessive. This could be due to a combination of two things—to the higher centre of gravity and roll centre of the ute-type chassis, and to the extra weight-loading imposed upon the car-type shocks. We think the fitting of heavier-pressure valves in the shock-absorbers would help here, and possibly the inclusion of an anti-roll bar would cure the trouble. The ride, likewise, is firmer than that of the car, by reason of the heavier ute springing. It is comfortable enough though, and we were unable to make the Minor's suspension bottom in many miles of beyond-normal usage.

Vision is exceptionally good all round, and the very good steering lock and general manoeuvrability of the little vehicle make traffic work a pleasure. We noted too that the strong quarter-bumpers at the back, and the strong reinforced tailgate between them, make the Minor stationwagon almost Sydney-traffic-proof against assaults from the rear.

Worthwhile point!

Good Puller . . .

Climbing ability on all gears was beyond the average, even when fairly heavily laden, and no hill we encountered forced us to engage second gear more than briefly. It was also a pleasure, in a Morris Minor, to be able to sit in line in Sunday traffic awaiting an opportunity to pounce—and then duck through, past, and away in a trice *without needing to change to a lower gear!*

We found that the 'wagon attracted a great deal of attention on the road—particularly from Morris saloon owners, most of whom were under the impression we had imported it from England. On learning differently, most were impressed with the quality of the Australian finish—and they were even more impressed when they tried to follow us and found the same low-geared little 'wagon walking away from them!

Their turn came, of course, when a clear piece of road arrived, for the 'wagon seemed to be as flat as a biscuit at an indicated 63/64 m.p.h.—though we did see an indicated 75 on one occasion following a downhill wind-up. The speedometer was

CONTINUED ON PAGE 14

Dash is of well known Minor style; has twin glove boxes and central instruments. Low final gearing gives the 'waggon almost sporty performance on the road — although fuel economy is poorer than that of the saloon.

MINOR WITH POWERPLUS CONVERSION

Twin Carburettors, Raised Compression Ratio, New Exhaust, Higher Gearing

Standard exterior—the difference lies beneath the bonnet

WHEN the Morris range for 1957 was announced shortly before the Earls Court Show last year, a number of changes which keen Minor owners had hoped for were, in fact, incorporated. A higher power output was obtained by increasing the engine capacity from 803.5 c.c. to 948 c.c. and raising the compression ratio from 7.2 to 8.3 to 1. At the same time indium-flashed lead-bronze main and big-end bearings were fitted, the crank pins being of increased diameter, and gear ratios were raised.

The result was the transformation of an established favourite into a machine which, in its price class for family cars, is unrivalled for performance and handling qualities. For those less interested in such matters the Minor continues to provide very economical transport for up to four persons and a fair quantity of luggage—in short it is a dual-personality car.

Inevitably, with such a sound basis on which to work, means of increasing the Minor's performance still further have been sought, and we have recently had the opportunity of driving a Minor 1000 Powerplus conversion by the Wicliffe Motor Co., Ltd., whose head office is at Russell Street, Stroud.

Modifications consist of a cylinder head machined to give a compression ratio of 8.75 to 1, twin 20-degree semi-down-draught S.U. carburettors with separate Smiths air cleaners, mounted on separate induction stubs with a balance pipe, a new central exhaust branch with no hot-spot, a Burgess "straight through" silencer, special Terry valve springs and a crown wheel and pinion assembly giving an axle ratio of 4.2 to 1 (standard ratio is 4.5 to 1). The cost of this kit of components for Minor 1000 owners to fit themselves is £67 4s, inclusive of an exchange re-calibrated speedometer.

There is ample room in the engine compartment for the twin carburettor installation; a chromium-plated valve cover added to the impression of cleanliness and efficiency

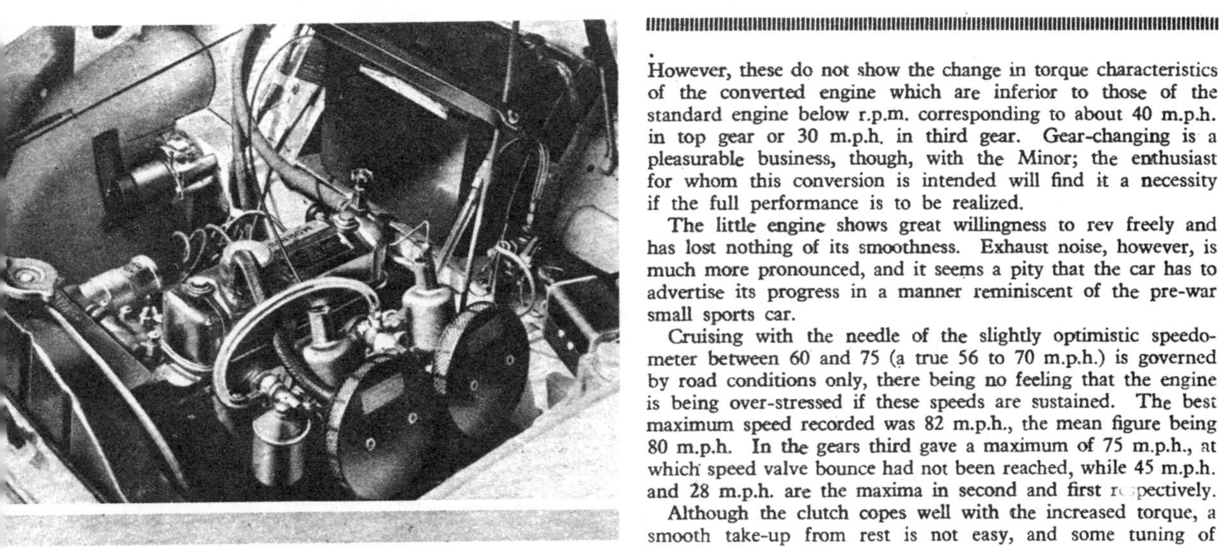

The first opportunity for a long journey in the Powerplus Minor came when Snetterton was visited for the July National meeting, the open stretches of A11 allowing the car to show its paces. The 76 miles from Hemel Hempstead to Thetford were covered at an average speed of 43 m.p.h., including a diversion among the serpentine lanes near Harpenden.

That acceleration is considerably improved can be seen from the table on this page, in which comparison is made with figures obtained during THE AUTOCAR road test of 14 September 1956.

ACCELERATION

	Powerplus conversion	Standard Minor
From rest through gears to:		
M.P.H.	sec	sec
30	5.6	6.8
50	13.7	18.8
60	20.7	31.3
70	30.5	—
Standing quarter mile	21.5	24.2
From constant speeds:		
20–40 m.p.h. in 2nd	5.7	—
30–50 m.p.h. in 3rd	9.1	11.7
40–60 m.p.h. in top	17.4	23.9

However, these do not show the change in torque characteristics of the converted engine which are inferior to those of the standard engine below r.p.m. corresponding to about 40 m.p.h. in top gear or 30 m.p.h. in third gear. Gear-changing is a pleasurable business, though, with the Minor; the enthusiast for whom this conversion is intended will find it a necessity if the full performance is to be realized.

The little engine shows great willingness to rev freely and has lost nothing of its smoothness. Exhaust noise, however, is much more pronounced, and it seems a pity that the car has to advertise its progress in a manner reminiscent of the pre-war small sports car.

Cruising with the needle of the slightly optimistic speedometer between 60 and 75 (a true 56 to 70 m.p.h.) is governed by road conditions only, there being no feeling that the engine is being over-stressed if these speeds are sustained. The best maximum speed recorded was 82 m.p.h., the mean figure being 80 m.p.h. In the gears third gave a maximum of 75 m.p.h., at which speed valve bounce had not been reached, while 45 m.p.h. and 28 m.p.h. are the maxima in second and first respectively.

Although the clutch copes well with the increased torque, a smooth take-up from rest is not easy, and some tuning of

CONTINUED ON PAGE 66

We compare 

WE took a perfectly standard Morris Minor and drove it hard. The result was 33.8 m.p.g.

A month later, we took a twin-carburettor Minor to the same test strip and drove it hard. This time the mileage worked out at 34.3 m.p.g.

Finally, a week or so ago, we took out a distinctly hot Minor. After repeating similar tests we measured the fuel consumption. It worked out at 44.9 m.p.g!

The stock Minor was supplied by the British Motor Corporation and was in excellent order. The other two came from Automotive Carburettors, Sydney, where the modifications had been done under the supervision of Wally Warneford.

The warm Minor was far livelier than the stock version and the third (and most economical) was livelier still. Does it make sense to get more power *and* more economy from a warm-up? It certainly does. The secret lies in that magic phrase — volumetric efficiency.

By improving the breathing of the engine, we found it was possible to extract more work from each "mouthful" of gas. With improved efficiency came more power and lighter throttle openings.

Everyone knows this is true in theory. But it's not often you get a chance to balance theory with practice. As our test comparison panel shows, the successful modifications to the Minor engine brought with them an appreciable all-round improvement.

Notice that the improvements were triple. Fuel consumption was reduced, power lifted and the top gear flexibility greatly increased.

So the modifications were not just simple, hop-up changes. It is comparatively easy to work over an engine so as to raise power. In many cases it's done by shifting the point of maximum torque further up the r.p.m. scale, so that the combination of torque-times-engine-speed gives increased b.h.p.

More power sometimes means poor idling, less flexibility and higher fuel consumption. In the Warneford jobs we tried, the engines were treated so that idling remained normal, flexibility was improved and fuel consumption came down.

Let's take a second look at the performance figures. Notice that the stock Minor took 14.9 sec. to pick up from 20 m.p.h. to 40 m.p.h. in top gear. The very warm Minor slashed the time by 3.1 sec. — an improvement of 21.5 per cent.

The hottest Minor also flashed over the standing quarter in 2.8 seconds less than the stock car — 11.2 per cent better.

Fuel consumption sank by 11.1 m.p.g. —an improvement of no less than 33 per cent.

All this was achieved by modifying the original engine and adding a second carburettor. How come? Because the final version was able to breathe more readily than the factory car.

Before looking into the whys and wherefores, it is of interest to see what changes were made to achieve the results.

In the second car, which was moderately warm by most standards, Wally Warneford made no attempt at serious modification. He merely made the most of the stock engine. To improve the breathing, he added a second 1¼-in. S.U. carburettor with a special manifold, appropriate fuel lines and throttle linkages.

The car was also equipped with a straight through exhaust, an exhaust elbow and a sports coil. The only other modification was a port and polish job.

The exhaust elbow may call for some explanation. Normally, this B.M.C. 948-c.c. unit has provision

Twin carbs, big valves, improved mixture flow are key features of fully developed Minor engine. Wal Warneford's son Peter races this car often.

Minors

Stock car-- Economical

Warm car-- Thrifty

Hot car-- Miserly

WHY?

for a hot-spot in the manifold. This limits the flow of exhaust gas from cylinders two and three because of the by-pass required to warm the hot-spot.

Several overseas firms have developed an exhaust manifold adaptor which provides a wider area of cross section for the exhaust flow. This is fitted in place of the hot-spot. Automotive carburettors make their own version of the elbow and it is now on sale, complete with studs and gasket.

The modifications described so far amount to nothing much more than the fitting of a twin-carburettor kit to an otherwise standard engine. The effect this had on the car's performance can be seen on our comparison page.

To get even more power from the same engine in the third car, Automotive Carburettors took the

PEDR DAVIS

Interesting detail touch on hottest car we tried was this supplementary dash panel. Instruments (revs., temp., fuel, oil pressure) are ex-aircraft. They look terrific on black panel.

modifications still further. Bigger valves were fitted, together with stronger valve springs. All ports were carefully matched.

The new valves started life as Bedford truck units and they were chosen because the valve heads were close to the size deemed best for the Minor engine and also because the valve stems — 5/16-in. in diameter — were suitable for the heavier springs. These in turn came from a Triumph motor cycle.

Performance figures indicate that the power output from the engine is around 50 b.h.p., an appreciable boost from a unit intended to develop 37 b.h.p. at 4,800 r.p.m.

An interesting point is that this has been achieved without modifying the camshaft contours. It is also a compliment to the Morris factory, in that no changes were necessary in the lower half of the engine. Both the oil system and cooling departments are stock. Apart from fitting a sports-type coil, the only change in the ignition system was the fitting of plugs one stage cooler.

Does all this mean that someone has managed to beat the fac-

WE COMPARE THREE MINORS . . .

" . . . To get more power, more m.p.g., improved flexibility and an even tickover all at once means tuning of the highest order . . ."

tory at their own game?

Not at all.

Manufacturers know all about engine tuning. In many cases they deliberately keep the power output down, partly because it is cheaper from a production point of view and mostly because experience has shown that a de-tuned engine lasts longer and gives fewer service faults. B.M.C. have been familiar with the effect of gas flow on their engine designs for many years. Since 1945 at least they have retained the services of Harry Weslake, well-known British cylinder head consultant.

Remember the engine fitted in the 1946 Austin 16 and later in the A 70, Atlantic and early Austin-Healey?

Few sportsmen know that this power unit was developed at the request of the British Government during the war for use in imported Jeeps. The authorities feared that the German blockade would cause the army to starve of engines and spare parts. As an emergency measure, they instructed Austin's design department to develop a suitable substitute.

There was no need to put the engine into production — American parts continued to reach England, and the Austin power unit remained in wraps until the dawn of peace. Then it was resurrected and applied to the new 16 h.p. car.

Because the basic engine had been designed as a relatively low efficiency unit, boss man Leonard Lord (as he was then) instructed the design department to send a prototype to Harry Weslake for gas flow study. With the aid of special equipment, Weslake developed a new cylinder head, a modified combustion chamber and new valves. That way he boosted power output from 54 b.h.p. to 68 b.h.p. Later, output was raised again to 88 b.h.p., using twin S.U.s.

We raise this point to stress that volumetric efficiency problems are nothing new to B.M.C. (nor to other major companies). But it is surprising that they fail to take full advantage of the power increases they could get with progressive design. As we pointed out before, warming an engine to give more power at the expense of other characteristics is relatively easy. To get more power, more m.p.g., improved flexibility and an even tickover means tuning of the highest order.

COMPARISON CHART

(Note: All three cars were tested under similar weather conditions).

	Stock Minor	Warm Minor	Hot Minor
Standing quarter-mile	25.0 sec.	23.0 sec.	22.2 sec.
Maximum Speed (two way average)	67.2 m.p.h.	72.0 m.p.h.	78.0 m.p.h.
Fuel Economy (hard driving)	33.8 m.p.g.	34.3 m.p.g.	44.9 m.p.g.
ACCELERATION:			
0-20 m.p.h.	5.2 sec.	3.1 sec.	3.0 sec.
0-30 m.p.h.	7.7 sec.	6.2 sec.	5.9 sec.
0-40 m.p.h.	13.4 sec.	10.4 sec.	8.9 sec.
0-50 m.p.h.	18.4 sec.	16.2 sec.	14.3 sec.
0-60 m.p.h.	32.2 sec.	25.0 sec.	22.2 sec.
TOP GEAR ACCELERATION:			
10-30 m.p.h.	14.4 sec.	12.0 sec.	11.7 sec.
20-40 m.p.h.	14.9 sec.	12.6 sec.	11.8 sec.
30-50 m.p.h.	17.1 sec.	14.0 sec.	13.4 sec.
40-60 m.p.h.	23.7 sec.	22.0 sec.	18.1 sec.
SPEED IN GEARS (valve bounce):			
I	22.0 m.p.h.	22.2 m.p.h.	24.0 m.p.h.
II	35.0 m.p.h.	35.0 m.p.h.	38.0 m.p.h.
III	57.0 m.p.h.	58.0 m.p.h.	62.0 m.p.h.

That's just what Wally Warneford has done with the Minor engine.

His first step was to increase volumetric efficiency. This automatically led to twin carburettors as one way of ensuring that each cylinder got the correct amount of mixture with every induction stroke.

Unless the gases have a ready flow to and from the cylinders it will be impossible to develop maximum brake mean effective pressure. This means that the shape of the combustion chamber, the positioning of the valves, their sizes, lift and angle and the positioning of the spark plugs areall critical. With stock engines, some of these factors cannot be readily changed.

Improving the gas flow is easy meat. When a gas flows through a passage there are always losses in energy which mean an equivalent loss in pressure. The losses are due to friction against the walls of the channel and also to friction between adjacent layers of gas moving at different velocities. Any turbulence with the gas will result in a pressure loss, though maximum turbulence in the combustion chamber itself is most desirable.

All pressure losses reflect in an overall drop in volumetric efficiency. In other words, the weight per c.c. of mixture drawn into the cylinders is below par.

Improving power output by increasing the volumetric efficiency is an automatic way of ensuring a reduction in fuel mileage. How did Automotive Carburettors also improve flexibility?

To a certain extent this followed the boost in volumetric efficiency. But credit is also due to the fact that the manifold and carburettor sizes were right for the job and the needles critically selected.

Carburettor sizes are far more important than most amateurs realise. At first sight, you may get the idea that as much air should be drawn into the carburettor as possible. But achieving this with an over-generous throat size (i.e., fitting 1½ in. S.Us. when 1¼ in. would do) is the first step to failure.

Low speed torque and general engine flexibility depend on a correct air flow/fuel mixture all through the range. At any speed only a given amount of air can be sucked into the engine. Since the velocity of the column of air is proportional to its volume, it follows that air drawn through a small orifice will travel at high speed while that drawn through a larger area will be slower.

If the velocity of air at low speeds is too slow, then the amount of fuel picked up will be insufficient.

In this respect alone the Warneford conversions should serve as an example to the aspiring hot-up enthusiast. #

ONE IN A MILLION

First British car ever to reach a million

Already a million owners know—you discover something great when you drive a Morris Minor 1000. You discover how spacious, lively and big-car-minded an economical car can be. You discover why the Minor is Britain's most successful car—a brilliant 'classic' of modern motoring. And finally you discover this: no other car would suit you and your family half so well! Meet the Minor 1000 personally, at your Morris dealer. Prove for yourself: a million 'Minor' owners *must be* right.

✱ Outstanding 'Minor' successes in tests of performance and sustained reliability include: the only car to complete 10,000 miles non-stop (Goodwood in 1952); first-ever car to complete 25,000 at an average of 60 m.p.h. on German Autobahnen in 1956; and the amazing Rome-London ride (1,067 miles) completed in a day at an average speed of 45 m.p.h.

Where is the oldest post-war 'Minor'? *We want to find it! To mark the millionth 'Minor', we will exchange a new 'Minor 1000' for the 'Minor' with the earliest post-war chassis number. The only stipulation is that the car must have completed 100,000 miles. If you think your 'Minor' qualifies, check that chassis number now and send details to Morris Motors Ltd., Cowley, Oxford.*

I'm going to have a

"QUALITY FIRST"

BMC

Prices from £416 (plus £174.9.2 Purchase Tax) Twelve Months' Warranty and backed by B.M.C. Service, the most comprehensive in Europe.

MORRIS MOTORS LIMITED, COWLEY, OXFORD · NUFFIELD EXPORTS LIMITED, COWLEY, OXFORD AND AT 41/46 PICCADILLY, LONDON, W.1.

Morris Minor 1000

2-DOOR DE LUXE

In its exterior appearance the Minor still shows few differences from the original functional body style. Overriders are included in the de luxe specification

WITH the introduction of new models and the replacement of contemporary designs, it is easy to forget how remarkably good is the Morris Minor 1000 as a current small car choice. Over 12 years have passed since the original Series MM Minor made its triumphant debut, yet the little-altered body style is still efficient and practical for today's needs. With the 1957 changes in the engine and gearbox and progressive detail improvements, the car has long reached the stage where few buying in this class would ever regret their choice of it. We are glad of the opportunity, in response to requests from many readers, to reassess its attributes. (The Minor 1000 was tested by *The Autocar* of 14 December 1956.)

Experience has proved completely the success of the B.M.C. Series A engine formula, in which an 8·3 to 1 compression ratio is used, in conjunction with large-diameter lead-bronze bearings. The engine has a very reasonable working life when treated intelligently, and it proves an admirable performer on the road. At low speeds it is extremely flexible, and the car will pick up speed without snatch from as little as 13 m.p.h. in top gear.

Throughout the speed range, which extends right up to a readily usable 5,000 r.p.m., the power unit remains unobtrusively smooth, and there are no vibration periods. The two-way average maximum speed of the car is just over 70 m.p.h., but in favourable conditions the engine will happily run up to much higher speeds, to such extent that the speedometer could well be marked above its 80 m.p.h. limit. With the optimism of 6 m.p.h. in the instrument on the test car at the upper reading, it was possible on a downhill straight to take the needle right off the scale and beyond the ignition warning light. For sustained cruising an indicated 70 m.p.h. (4,350 r.p.m.) appears to be well within the performance ability of the Minor. The engine is a reliable first-time starter, but even in mild weather the choke is needed for the first start of the day, and for the first few minutes the control has to be kept open one or two notches to avoid hesitation.

Engine noise is never excessive, and although a fair amount of "buzz" is audible when revving up in the indirect gears, it mingles with the prominent gear whine—particularly in third—and is not obtrusive. Not at any speed attainable in top gear does the engine sound as though it is revving dangerously fast. There is a crisp, healthy exhaust note, which is resonant on the overrun.

One of the features which make the car a delight to drive is the excellent four-speed gearbox with remote-control central change. The lever is well placed within easy, natural reach of the driver's left hand, the change being precise and extremely smooth in action. There is no notchiness, and little resistance is felt as the gears are engaged; the lever is not spring-loaded to either plane in neutral. There is no synchromesh on the bottom ratio; on the test car—which had covered some 10,000 miles—the synchromesh was not powerful on second gear, and could be beaten also in rapid changes into third.

All the ratios are well matched to the engine characteristics. Second seems a little low, but it has the advantage that starts from rest can be made in this gear. Normally, however, each ratio is used to the full when moving away from a standstill, second gear providing a readily usable 25 m.p.h., and a maximum of comfortably over 30 m.p.h. In third gear, 50 m.p.h. is well within the available range, and on the open road this gives the Minor a lively turn of acceleration for overtaking. The power can be felt to develop strongly at 30 m.p.h. in third, and the car accelerates over the all-important 30-50 m.p.h. speed gap in 12·6sec. At a standstill, bottom gear is sometimes reluctant to engage, but there is no difficulty in selecting it if the clutch is released for a second with the lever in neutral. Reverse gear position, which is next to that of top gear, is protected by a safety spring strong enough in normal driving to prevent accidental engagement, but not so powerful as to obstruct the selection of reverse when required.

A flexibly mounted rod serves as the link in the clutch-operating mechanism, and the action of the withdrawal is smooth and extremely light. Take-up is smooth, even if the pedal is released abruptly with the engine revving fast.

One of the most striking things noticed on taking over the Minor for the first time is the precision and lightness of the steering. Even at manœuvring speeds it remains extremely light to operate, and the Minor is simplicity itself to park, thanks to its quite compact turning circle, and the

The roomy luggage locker has a wooden floor, with the spare wheel sensibly stowed underneath. The toolkit includes a jack, wheel nut spanner, screwdriver and hand pump, and there is a starting handle

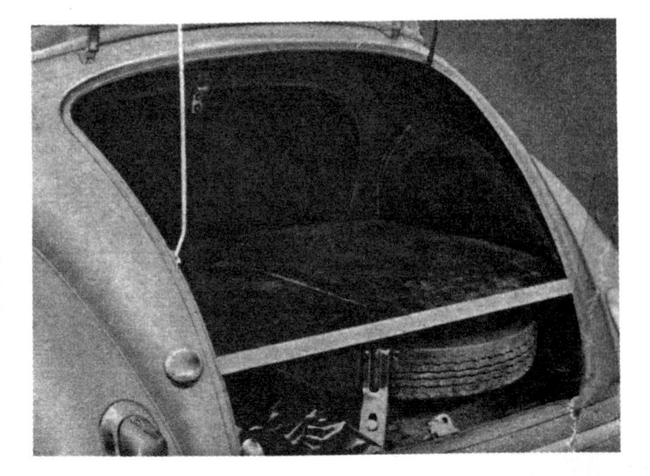

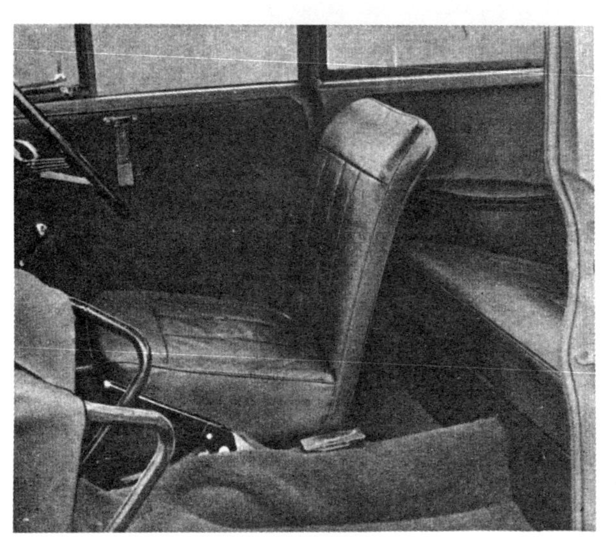

Access to the rear seat is easiest through the left door, because not only does the front passenger seat pivot forward on its leading edge but its squab also folds down out of the way. Right: A front ashtray is provided below the facia, and there is another for rear seat passengers on the propeller shaft tunnel. Twin glove compartments (but with unlockable lids) and a full width parcels shelf are provided. Washable plastic lines the roof

ability to change the lock rapidly within a comparatively small forward or rearward movement of the car. In this respect also, the good all-round visibility is an advantage, but the short snub-nose bonnet obscures the left front wing, and some drivers may have difficulty in judging the position of the car's extremities.

Light steering is achieved sometimes only at the expense of lack of precision, but with the Minor's rack-and-pinion layout the driver has the best of both worlds. Standing beside the car and watching the wheels while the steering is moved, it is noticed that the tiniest turn of the control is transmitted to the road wheels; there is no detectable lost motion, and this is amply evident on the road. Imperceptible steering corrections can control the course of the car, and although the Minor's good directional stability is much affected by a cross wind, it is again easy to maintain a very straight course because of the ready response to the steering. In spite of this high degree of precision there is surprisingly little kick-back through the steering over rough roads.

Reasonably stiff suspension gives the car a notably stable ride, with little pitching, and quite adequate absorption of average surface irregularities. Only over a severe bump or pothole is a minor suspension weakness brought to light —a tendency for the transmission of a sharp jolt and rebound rattle from the front suspension. On rough, unmade tracks there is a fair amount of firm vertical movement of the car, but bouncing at front or rear is strongly controlled

For extra luggage capacity the rear seat squab may be folded forward, extending the stowage space into the rear interior of the car. With care, the boot lid can be pushed over-centre to rest against the panelling

by the dampers, and the resilient rubber bump stops eliminate any harsh bottoming even over severe colonial conditions. On certain main road surfaces some road "rumble" is transmitted to the body from the wheels, and at the top end of the speed range—above 70 m.p.h.—there is a considerable road roar, although engine and wind noises remain at a low level. At speed also, a strong draught found its way past the sealing of the left door.

Very little body roll occurs on corners, but the effect of weight transference produces a tendency for the inside rear wheel to spin or sometimes for a severe and rather alarming axle tramp to develop. This characteristic is well known, and proprietary modifications are available to restrict it. In normal driving, of course, the fault does not occur, and many owners may never be troubled by it, since it appears only when accelerating hard out of a sharp corner.

Slight understeer is detectable in the behaviour of the unladen Minor on corners, but the balance as a whole is not far from the neutral condition, and the car is never difficult to control. To the inexperienced it gives confidence, and on wet, slippery roads it again remains easy to handle. The back of the car seldom slides unless a skid is provoked deliberately, and if this occurs it may be corrected without difficulty, so long as the precision and lightness of the steering are remembered, and the tendency to over-correct is avoided. The car has to be cornered hard indeed before tyre squeal is heard.

These good road manners are backed up by efficient brakes. Pedal pressures needed for normal braking from cruising speeds are not unduly high, but quite a firm push is needed for anything approaching an emergency stop. When violent braking is needed the power for it is there, and the car dips slightly at the front and stops "all square" with reassuring smoothness. Hard braking can still be used on wet roads with little danger of the wheels locking. Fade did not occur during the test. Placed conveniently between the individual front seats is the pull-up control for the hand brake. Its action enables real power to be exerted, and the hand brake held the car securely on a 1-in-3 test hill.

Like the hand brake, all other controls are well placed where they fall within easy reach of the driver, the only possible exception being the facia switch for the head and side lamps, which is not easily found when it is needed momentarily for flashing the lamps. Just below the steering wheel rim is the indicators control, and we are pleased to note two improvements which we advocated in our earlier test of the Minor 1000. The first is that the indicators have now been made self-cancelling, and the second,

Morris Minor 1000...

Stop and tail lamps are combined with the reflectors in one unit on each side at the back of the car, and semaphore trafficators are retained

that the horn switch is divorced from the trafficator control. The horn has a deep and penetrating single note, and is operated by a button in the centre of the steering wheel.

Swivelling quarter-lights are fitted in the doors, but there is no anti-thief device on their latches. As the door handle is just below they give access to the car if they are forced, but to prevent this it is a simple matter to reverse the catch plate, making it difficult or impossible to open the window from outside.

Mounted centrally in the facia, the speedometer is obscured from view by the driver's left hand in the normal "ten-to-two" position on the steering wheel. Set in the instrument are warning lights for oil pressure, ignition, and the head lamp main beams, and a tolerably accurate gauge for the 6½-gallon fuel tank.

The Minor's fuel capacity has come up in three stages, from 5 gallons, through 5¾ to the present figure of 6½ gallons, which now gives the car a useful range between refuelling, thanks to the commendable petrol economy. Repeated spot checks in varying traffic conditions with hard as well as restrained driving did not return more than 40 m.p.g., yet the consumption was seldom far removed from this figure. The overall consumption of 34·7 m.p.g. represents a great deal of the sort of hard driving which the security, liveliness and ease of control of the Minor invite, and is a fair minimum m.p.g. figure which most owners should easily exceed.

The self-parking windscreen wipers are pivoted from the extremities of the windscreen—a hangover from the original divided screen version of the Minor. At the base of the screen the blades come together, but as they separate a central "vee" area is left unswept. Also, the travel of the wiper blades on each side ends some 2in from the edge, resulting in additional unwiped areas. There is much the owner could do to improve this, by fitting longer blades and phasing the wipers to overlap at the bottom, but there seems no reason why the manufacturers should not get it right in the first place.

Interior comfort of the Minor is commendably good for a small car. The bucket seats provide adequate support, and are softly sprung, while rear-seat comfort for two adults is above average in this class of car. Ample legroom is provided in the rear compartment, but at the expense of accommodation in the front. On the test car the driving seat was on its rearmost adjustment for all drivers, and those with long legs would certainly need to set the seat farther back to avoid sitting with knees up, and sharply bent. The passenger seat is not adjustable for reach. Under the shaped but rather loosely fitted carpets are seen additional mounting holes (with captive nuts beneath) in which the seat may be repositioned 1in. to the rear, and we have seen a Minor on which new holes for the seat had been drilled even farther back.

In the normal position the steering wheel is at convenient arm's length from the driver, and the slight offset of the column to the left is scarcely noticed. The spacing of the pedals is just about sufficient for a driver wearing broad-soled shoes to be in little risk of overlapping from one pedal to its neighbour, yet "heel-and-toe" gear changes (using the side of the foot on the accelerator) are conveniently easy.

When a long or particularly cumbersome item of luggage has to be carried, it is the work of a moment to release a Lift-the-Dot fastening in the boot and push the rear-seat squab forward on its sliding hinge. As there is no rigid division between the luggage locker and the car's interior, unusually generous accommodation is thus available when no rear-seat passengers are carried. Indeed, we know of one Minor owner who carried a full-size motor cycle (with the front wheel removed) in the luggage compartment in this way; there are others who have found their way through the unlocked boot into the car when they have mislaid the door key. A worthwhile improvement would be the provision of a self-sustaining strut or springs to hold the boot lid open. The existing strut now clips into a rubber socket and is easier to release than on the earlier models where a metal clip was provided; but opening the boot lid and fixing its stay in position is still a job requiring two free hands.

The two-door de luxe Minor—previously we tested a four-door car—shows a saving of £40 (total) in return for the limited inconvenience of not having rear doors. An advantage of the two-door car is that the openings are slightly larger, making access to the front compartment easier than on the four-door. Passengers find no difficulty in climbing into the back of the car from either side. The weak point is ventilation, all windows to the rear of the doors being fixed.

Included in the de-luxe specification is a fresh-air heater

In the cavernous under-bonnet compartment the engine is almost dwarfed, and accessibility is particularly good. A modern type of air filter to funnel cool air from the side of the radiator grille is now fitted

with a tap under the bonnet for the hot-water supply. With this turned off, the heater may be used to provide a supply of cool air to the interior. Alternatively, in colder conditions the heater proves powerfully effective, but it is difficult to set it to deliver small quantities of warm air. Other de luxe features are leather seat upholstery, an extra sun visor (for the passenger) and bumper over-riders, which seem fully to justify the extra cost of £29 on the total price. On the standard car the seats are upholstered in Vynide.

The power of the standard double-dipping head lamps covers amply the performance range of the Minor. No reversing light is fitted nor is there provision for one. A bright interior light is switched on automatically when either door is opened, and there is also an over-riding switch.

Owners who carry out their own vehicle maintenance will rejoice in the excellent accessibility of all under-bonnet components; and will be glad to know that there are only ten 1,000-mile grease points, including the hand-brake cables.

In every way the Minor is a thoroughly practical small car which has the special qualification of being immensely enjoyable and safe to drive. Eventually, of course, the day will arrive when its production run will come to an end, and the Minor will go down as one of the cars which has made motoring history. In the meantime, it is still to be looked upon as a strong contender in the present-day battle for the "best small car on the market."

MORRIS MINOR 1000 2-DOOR DE LUXE

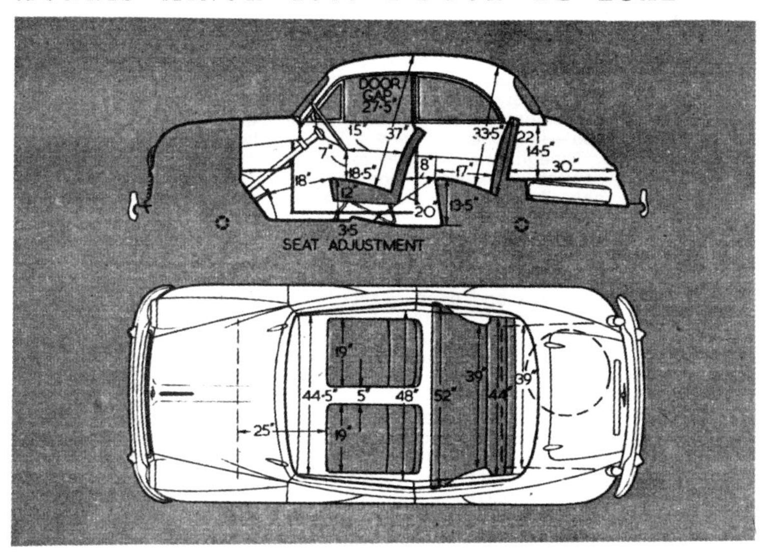

Scale ¼in. to 1ft. Driving seat in central position. Cushions uncompressed.

PERFORMANCE

ACCELERATION TIMES (mean):

Speed range, Gear Ratios and Time in Sec.

M.p.h.	4·55 to 1	6·42 to 1	10·80 to 1	16·51 to 1
10—30	—	10·3	6·1	—
20—40	14·6	9·9	—	—
30—50	17·9	12·6	—	—
40—60	26·1	23·9	—	—

From rest through gears to:

30 m.p.h.	7·2 sec.
40 "	12·0 "
50 "	18·8 "
60 "	32·6 "

Standing quarter mile 23·5 sec.

MAXIMUM SPEEDS ON GEARS:

Gear		m.p.h.	k.p.h.
Top	**(mean)**	73·1	117·7
	(best)	76	122·3
3rd		60	96·6
2nd		34	54·7
1st		22	35·4

TRACTIVE EFFORT (by Tapley meter):

	Pull (lb per ton)	Equivalent gradient
Top	163	1 in 13·7
Third	240	1 in 9·3
Second	353	1 in 6·2

BRAKES (at 30 m.p.h. in neutral):

Pedal load in lb	Retardation	Equiv. stopping distance in ft
50	0·30g	100
75	0·48g	63
100	0·74g	41
120	0·81g	37

FUEL CONSUMPTION (at steady speeds in top gear):

30 m.p.h.	50·0 m.p.g.
40 "	47·6 "
50 "	40·0 "
60 "	34·4 "

Overall fuel consumption for 1,177 miles, 34·7 m.p.g. (8·1 litres per 100 km.).

Approximate normal range 31-39 m.p.g. (9·2-7·2 litres per 100 km.).

Fuel: Premium grade.

TEST CONDITIONS: Overcast, showers, 10-15 m.p.h. breeze.

Air temperature, 65 deg. F.

Model described in *The Autocar* of 12 October, 1956.

STEERING: Turning circle:

Between kerbs, L, 32ft 9in., R 31ft 7in.

Between walls, L 34ft 2in., R, 33ft 0in.

Turns of steering wheel from lock to lock, 2·5.

SPEEDOMETER CORRECTION (m.p.h.)

Car Speedometer:	10	20	30	40	50	60	70	80
True speed:	9	19	29	39	49	57	66	74

DATA

PRICE (basic), with two-door de luxe body, **£436.**

British purchase tax, **£182 15s 10d.**

Total (in Great Britain), **£618 15s 10d.**

Extras: Radio **£27 plus £11 5s,** purchase tax.

ENGINE: Capacity, 948 c.c. (57·8 cu. in.).

Number of cylinders, 4.

Bore and stroke, 62·9 × 76·2mm (2·48 × 2·99in.).

Valve gear, o.h.v., pushrods.

Compression ratio, 8·3 to 1.

B.h.p. 37 (net) at 4,750 r.p.m. (b.h.p. per ton laden, 39·1).

Torque, 50 lb ft at 2,500 r.p.m.

M.p.h. per 1,000 r.p.m. in top gear, 15·2.

WEIGHT: (With 5gals fuel) 15·9cwt (1,785 lb).

Weight distribution (per cent): F, 56; R, 44.

Laden as tested, 18·9cwt (2,121lb).

Lb per c.c. (laden), 2·2.

BRAKES: Type, Lockheed hydraulic.

Drum dimensions: F and R, 7in. diameter; 1·22in. wide.

Swept area: 107 sq. in. (113 sq. in. per ton laden).

TYRES: 5·00 × 14in. Dunlop Gold Seal

Pressures (p.s.i.): F, 22; R, 22 all condition

TANK CAPACITY: 6·5 Imperial gallons.

Oil sump, 6·5 pints.

Cooling system, 9·75 pints (plus 1 pint if heater fitted).

DIMENSIONS: Wheelbase, 7ft 2in.

Track: F, 4ft 2·6in.; R, 4ft 2·6in.

Length (overall), 12ft 4in.

Width, 5ft 1in.

Height, 5ft 0in.

Ground clearance, 6·75in.

Frontal area, 18·5 sq. ft. (approximately).

Capacity of luggage space, 7 cu. ft. (approx).

ELECTRICAL SYSTEM: 12-volt; 43 ampère-hour battery.

Head lamps: Double dip; 42-36 watt bulbs.

SUSPENSION: Front, independent, torsion bars and wishbones.

Rear, live axle and half-elliptic leaf springs.

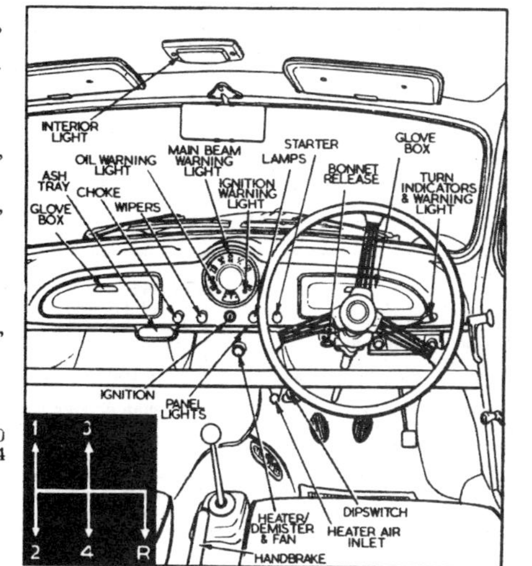

MORRIS MINOR

Price £718, and £43 extra for 4 doors.

FOR the type of man who doesn't know anything about motoring but knows exactly what he wants, the Morris 1000 is one of the rare cars that can give it him. Here is the most completely disinfested thing on wheels, with not a living bug left after all those twenty-one years in production. A car that can give such unfalteringly trouble-free service can no longer be said to have faults; they are now merely characteristic foibles, like panama hats or walrus moustaches, and very nearly as reverend.

The frog-eyed, high-bonneted, tumble-home Minor has let the world of fashion pass it serenely by. This is not altogether a bad thing: the small tyre sections are compensated by large diameters, making the load-bearing capacity of those old cross-plies greater than you might estimate. A deeply rounded roof and other bulbous body panels (the wings are detachable) may constrict interior space, but on the other hand they make for very strong arches and shell structures. *Looking* dowdy, fat and 'fortyish may not matter; but there would be no scope on the roads of the 'seventies for a car with the *performance* standards of the 'forties, and this is where the Minor scores . . . even over the VW. That it continues to rank among the more sprightly of small cars is explained by two things: first, the regular and frequent upgrading of its innards—from a relic related to a pre-war side-valve engine, to the present one, which is really an 1100.

The other thing is that the original design concept was so precocious that the car is still capable of handling the extra power and performance of today. In fact it was originally built to handle more, being conceived as a 1½-litre affair with a brand new flat-four engine potentially adaptable for front-wheel-drive. Only now, in the autumn of its life, is it getting the sort of power that was envisaged by Issigonis in the first place. Maybe this was as well, in the light of later BMC disasters! In the intervening years the Minor has been able to build an enviable reputation for road-worthiness as well as durability. What with perfect balance, wide track, and a well-conceived torsion bar front suspension that uncommonly combines longitudinal compliance with precise long-lasting screw-thread bearings for the steering pivots, the Minor can still be cornered hard and driven energetically, even by current standards. Funny to recall that when it was young people were saying that the car was *too* wide, too low, and too anonymous-looking!

Issigonis designed it while still keen on racing, as everyone knows, and designed it well. Good design does not date. There isn't a safer car on the road (ask any insurance man); or an easier one to break into (via the boot), or a more considerate one on tyres (ask Dunlop).

The pedals emerge from the floor to meet the leg muscles in the manner that doctors think they should, but the steering column has that offset awkwardness which so many dislike.

Nevertheless in no other respect can you expect it to match up. The seats have precious little adjustment and a tall driver can be acutely uncomfortable. If you want ventilation you open a window. If you want to change down to bottom gear you double-clutch. Crash safety padding? Phooey. Cranking handle? Of course.

This is a car which you enter with decency, in which you sit up and take a high view, a long view, a less dazzle-prone view,

Weight 1,708 lbs. Tyres/Rims — 5.20 × 14 ins. (3"). Fuel — 6½ gals. En
Oil — 6½ pints. Coolant — 9¾ pints.

further round corners (over the hedges) and—if you consider the dubiety of the supposed improvements in all the cars about you—a *jaundiced* view.

Though the interior is a bit spartan, one wants for remarkably little. The merits of solid dependability are evident everywhere. The Minor came to power in the pre-skimp days of car-building, has enduring metal thickness and good quality hardware, with correctly inserted chassis reinforcements where they can't be seen to be believed, but work for the buyer nonetheless.

So the ubiquitous 1000 lives on. Every nut and bolt indelibly familiar to those who service them (and servicing is something you are asked to do religiously at 3,000 mile intervals). They haven't bothered to re-write manual yet, so those who t IT seriously must also ch water daily and go round tyres weekly, with a pump gauge in one hand and a h pick (to take out the flints nails of a by-gone age) in other.

Yet the public adores it will not let BMC stop mak them. Such buyers are not numerous as to justify the pense of retooling to make car more up-to-date, and in lies the secret of its surviva it *were* alterable, it would alre have gone the way of the that was intended to supplan

Most manufacturers give improvements in annual pe packets; in the case of the Mi BMC opened a credit acco which is not yet exhausted.

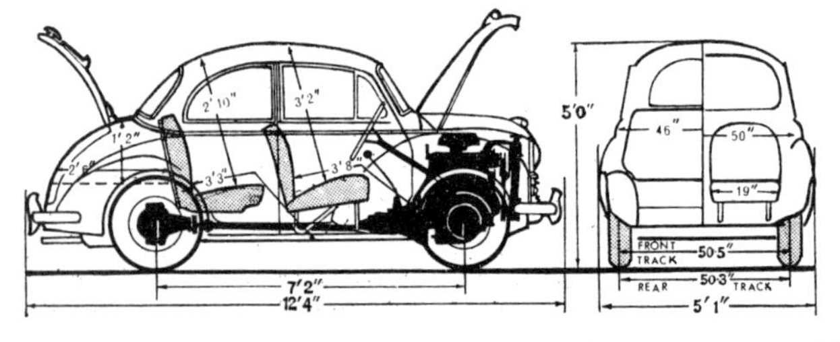

THE PLOT: General construction follows quality standard no longer acceptable in category. Water-cooled in-line front-mounted long-stroke 4-cylinder pushrod engine (3 mains) drives rear wheels via 4-speed gearbox and long unsupported propeller shaft. Change on floor. Rigid rear axle suspended from semi-elliptic leaf springs. Wishbones and longitudinal torsion bars at front. Chassisless body supported by much reinforcing. Rack and pinion steering. Two- or four-door.

FOR
- Virtually indestructible
- Good value. Reliable. Quiet
- Very safe, foolproof handling
- Low upkeep costs
- Straightforward. Easy to drive and maintain
- Well balanced. Light positive steering
- Maker-backed service

AGAINST
- Dated body style
- Non-synchromesh 1st
- No planned ventilation
- Poor parking lock
- Unrestrained rear axle hops on rough roads
- Busman driving position
- No anti-burst locks or padding
- Sealed beams

TECHNICAL HIGHLIGHT: That torsion bar front suspension remains a classical example, with long bars, carefully located lower wishbone assemblies, and a perfectly accurate steering plan.

DEPRECIATION is more closely allied to "datedness" than to the normal signs of wear and tear, but it is accepted that the overall fall in value is little retarded by repairs or renewals carried out by the owner before selling. It seems not to matter at all if the starter motor is a new one, or if the gearbox has been stripped down and rebuilt.

Therefore, it can be said that anything which avoids repair bills also saves money for the owner, particularly the first owner.

And here we find compelling reasons for being meticulous about regular maintenance, and for driving with restraint.

In the first half of its life of anywhere between 80,000 and 100,000 miles, a car reaches two stages. The first, at around 20,000 miles, when tyres have to be renewed, a new battery bought perhaps, and brakes attended to, new sparking plugs (second set?), and the odd bulb or two fitted. From the point of view of depreciation this is a bad time to change the car, provided it has been carefully looked after and driven.

THE second stage arrives when a major engine overhaul is required, together with more tyres, plugs and possibly a few ball ra in expensive-to-get-at places.

This, or slightly before it comes too obvious that repairs are needed, is the t to swap.

Stage Two mileage varies mendously between one driver another. The conscientious ow should be able to look forward 40,000 miles and maybe four y of trouble-free driving. The fel who hacks his car may laugh you for being fastidious, but he be well on the way to Stage T in his second car while you are on your first.

Regardless of our gloomy in duction the decision to chang car is governed more by its mile than by elapsed time, and n always take a nice calculation account: that of balancing the of repairs (if you hold on to against the loss on depreciation you don't).

Restrained driving is good more than the mechanical side saves a greater amount of pe than you realise, and can aln double the life of your tyres.

Still looking good. The registration letters, UUU, led to it being nicknamed "The Three Volunteers" early in its career.

'Thou good and faithful servant . . .'

115,000-mile Minor 1000—plus another 50,000 to come

How long can a normal small car be expected to last, given reasonable but not lavish attention? The one selected for this report started life as a company car and was then sold to a private individual who, although a technical journalist, is not on *Motor's* staff. He has used the car solely as family transport.—Ed.

by Bruce Main-Smith

THE Main-Smiths are not monied nor ever likely to be. So we keep off the bread line—for example, by having brake drums skimmed at 20s. a time instead of buying them new at £2. (It's salutary to realize that our labrador eats a brake drum a fortnight, if you see what I mean.) The November, 1957, Morris 1,000 4-door saloon, UUU 658, was my first real, modern, car. It came to me as a private citizen at a Warren Street price after doing sterling service in the hands of a Temple Press photographer—and they *use* cars! I got the Morris at 83,000 miles, when it was five years old, with a replacement Gold Seal engine in its prime and a carefully kept and detailed log. The present mileage is 115,000 and the service-exchange engine and gearbox have been the only expense of any consequence.

The car was originally run by Alf Long, one-time staff photographer, whose work has often illustrated the pages of *Motor*. Thanks to his log, and my records, we know where we are.

"Never on any occasion did it let me down," Alf wrote in answer to my questions. "I spent lots of time on Continental and Manx roads, about five International Six-Day Trials, five TTs and any number of car GPs. I never flogged it to death and seldom drove at maximum speed. Both engines were run-in—but not at 30 m.p.h. I never let it toil. Did my best to see it was serviced regularly, even when abroad. Used to get about 35,000 out of the tyres. Had some biffs with it: taxi hit one rear wing, American bus clouted the boot in Germany and an Austrian bus did it again at Garmisch-Partenkirchen. Left front savaged by hit-and-run van. Other drivers irresistibly attracted to it when it was stationary.

"Had one new battery and it was only when that was due that I had to start it on the handle. One burst top hose. No other bothers. All in all it was the nicest little car I have ever had. Its only disadvantage was its colour. Black gets so hot in the sun."

So wrote the first driver about his 83,000 miles of custody of a company car. Press schedules, green lanes, out in all weathers and seasons, salty sea crossings, fleet maintenance regularly and competently executed by the firm on its own premises, generally well-cared for in the best private-owner sense. Alf left the upholstery in fine condition. It still is. But I stabbed the headlining in two places when carrying an 8-ft. length of wood and this pvc has definitely yellowed with age.

The reasons for the replacement of the engine-gearbox at 61,000 miles are lost in the mists of time. The log simply notes the bare fact, together with a dynamo/light service. We can suppose fair wear and tear. The 948 c.c. Gold Seal replacement engine is now up to 54,000, and going well. Synchro is weak on top, almost nil on third but powerful on second. Bottom is crash as ever was. The clutch is a bit juddery and lacks bite on a brisk getaway. But it doesn't slip.

The Bendix starter either got some grit in it or the flywheel-gear-ring teeth are worn

Continued on the next page

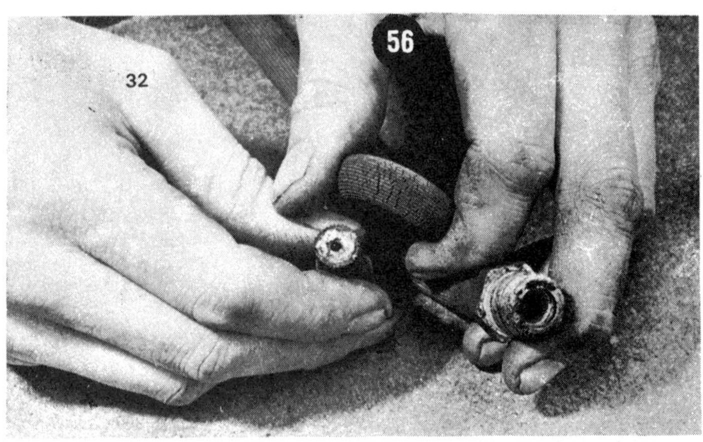

Brake hoses were replaced at 100,000 miles but should have been attended to much earlier—every three years is the norm.

The lower rear brake shoes show uneven wear after 5,000 miles, despite the addition of corrector springs.

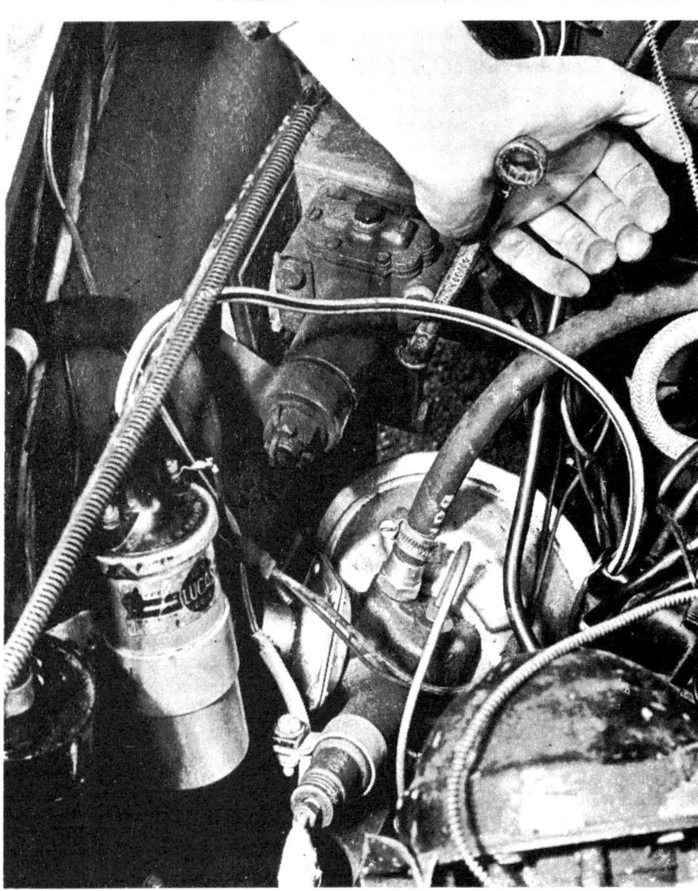

An odd feature is that the bolts retaining the front lever-type dampers need tightening every 500 miles because of the apparent compression of the damper body.

Thou good and faithful servant

Continued

because the original starter stuck for the fir time about three weeks ago. If it's the ri the gearbox may have to be removed soc and I can examine the clutch at the same tim Otherwise the box stays put, with the failir synchro, until I *have* to break the unit dow

The regulator is 1957 original and so the dynamo. The tailbearing has bee renewed once again, by me. Sundry brushe bulbs, contact-breaker points, one hea lamp shell (corroded beam-adjusting screws and two batteries have been the total electrical attention. As soon as I got th car I opted for a heavy-duty Exide, a batter with bags of weight to it—plenty of lea that's my criterion. It has been given regul doses of *distilled* water, as per the boo a higher-charge setting to the a.v.c., ar it's full of vigour three years later.

In Alf's time the car needed nothing oth than brake shoes, tyres, bulbs, plugs, exhau systems, one top hose and the Gold Seal un There is no record of minor attentions aut matically done during this period by th Temple Press garage, like door-lock adjus ment, wiper-blade renewal or the actu mileage when tyres were interchange between wheels.

As soon as I got the car I proceeded on th assumption I was taking it up to 150,00 miles at least and therefore renewed ear any minor components that would hav to be replaced in the foreseeable future. thought I might as well have the full use new material rather than send it to th breaker only part-worn! In my first 7,00 miles (i.e. by 90,000 total), I had put in on rear steel brake pipe—narrowly avoidi total brake failure by detecting in time chafi of the pipe on the back axle; a maste cylinder rubber kit; three brake whee

At a glance

Good

Handling
Handbrake power
Economy
Ease of Maintenance
Visibility
Sturdiness
Integral tow-bar mounts
Weather tightness
Ability to carry long and large objects
Natural, relaxed driving position
Open-door kerb-clearance
No sky probing with headlamps when laden at re

Mediocre

Gearbox*
Brakes*
Parcel shelf
Springing of back seat
Fuel tank capacity
Headlamp penetration*

Bad

Battery location
Ventilation
Claphands wipers*
Inadequate ground clearance
Small boot
Fuel filler
Noise from i.f.s.

Extras to taste. Additional fittings include vacuum gauge, water temperature gauge, tachometer, ammeter and two-speed screen wipers.

:ylinders; had the drums skimmed; fitted a "bunch of bananas" manifold and a new :xhaust system; a set of window felts, the dynamo tailbush plus a skimming of the commutator; a valve grind; cured a gearbox oil eak from the speedometer cable take-off; econditioned the speedometer head; re-ealed the driver's door; replaced the back window rubber; added a fresh-air kit to the Smiths recirculating heater; fitted new radia-or and heater hoses; and bought two new Dunlop C41s, placing them on the front.

Michelins were on the car when I got it. The trouble here is that Morris 1000 tyre wear is so slow that the sidewalls can erish before the treads wear away. I confirm Alf's 35,000 miles per cover and I'm no ussy-foot. All tyres wear equally. I am trongly against moving wheels to other orners and believe in re-balancing every uarter of a tyre's life.

Modern Minor winkers have been fitted. The old trafficator slots have been blanked with simple stainless steel plates.

Within a fortnight of taking over the car put the fan through the radiator by mis-udging the sump to gate-stop clearance in a Bournemouth gateway, when reversing; nis little fracas broke the gearbox trainer cable which takes the clutch pedal nrust. Every time I declutched or braked ard the fan made merry jingling noises. By dint of the white-of-egg-in-the-radiator ick (it seals it temporarily) I hopped from ight garage to garage for water. Farnham vas reached before I got overbold. As soon s the engine pinked, on the by-pass, I threw e clutch out but before I could hit the gnition key it locked solid. Dropping the ump and lifting the head to draw the rods, found No. 4 piston rather sorry for itself. : still is, 35,000 miles later. I suppose I hould have renewed the rings but between e 1,500-mile oil changes, I add one pint.

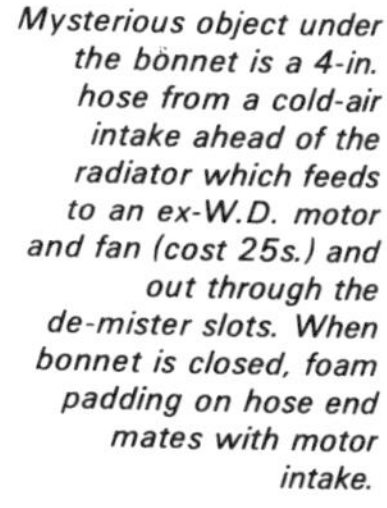

Mysterious object under the bonnet is a 4-in. hose from a cold-air intake ahead of the radiator which feeds to an ex-W.D. motor and fan (cost 25s.) and out through the de-mister slots. When bonnet is closed, foam padding on hose end mates with motor intake.

ootlid stay*
ackaxle tramp
ck of over-riders
maphore indicators*
cking system
ck of consideration for children
haust pipe and supporting-strap life
Current model improved in these respects

What It cost

	£	s.	d.
oad Fund Licence (3 x 4-months)	**19**	**4**	**0**
surance (Comprehensive, 50% discount), net	**12**	**11**	**0**
iginal cost of car (with tax)	**650**	**0**	**0**
esent resale value	**150**	**0**	**0**
epreciation	**500**	**0**	**0**
osts per 1,000 miles			
trol	**5**	**16**	**0**
l, filter	**–**	**15**	**6**
ugs and points	**–**	**2**	**6**
gine and gearbox wear	**1**	**0**	**0**
res	**–**	**17**	**0**
undries (battery, wiper blades, bulbs, brake shoes, anti-freeze, fan belts, etc.)	**–**	**10**	**0**
tal (exclusive of labour)	**9**	**1**	**0**
ost per mile of "direct" expenses: 2.16d./mile.			

This was the moment that the valve grind got done, otherwise I am an adherent of the "never decoke but drive it hard" school. It seemed prudent to put in new big-end shells. Further shells have since been fitted and Those Who Know have persuaded me that the A-type crankshaft will last for ever if the shells are changed every 12,000 miles: it only takes 90 minutes and as many six-pences. Anyway, every 25,000 the sump has to be dropped to clean the gauze oil strainer, a chore I am as careful about as I am never to exceed 6,000 miles on a filter element.

Realizing that the above policy is wide open to tut-tutting noises from the general direction of the British Motor Corporation, let me get my oar in first and say that it means I needn't expect to have the fag, the expense or the inconvenience of taking the car off the road for a crankshaft grind this side of a six-figure engine mileage. Call it "precautionary repairs".

To prevent a similar outcome of pilot error, the fan blades were removed and they've lain in the garage ever since. The engine didn't overheat in a 60-minute 6-mile Taunton holiday journey in high summer (I've since learnt the Taunton back-double). The car uses an 80°C thermostat in summer and an 86°C in winter. It has the number plate permanently fixed across the radiator where it renders the radiator blind somewhat redundant. Below the bumper, the plate acts as a bulldozer blade in snow until it folds under, ready to drop off.

The propshaft is the original (note for 1100 owners: its the thing that takes the drive to the rear wheels) and it hasn't even needed a Hardy-Spicer roller kit. The back axle is still silent and free from back-lash.

The windscreen wipers were changed over from the ghastly "clap-hands" arrangement to the current sort and while doing this I found that the outer cable (rack tube)

CONTINUED ON PAGE 68

150,000 MILES in a MORRIS MINOR 1000

By Michael Collier

WAY back in 1958 I conceived the desire to drive from London to Istanbul and back, and felt that my R-Type Bentley Continental was too valuable a machine in every way to hazard on this journey. In those days anything south and east of Belgrade was not well known from the motoring holiday angle and I wanted a suitable car, which would not be too precious.

After careful consideration of what was then available, I decided on a new Morris Minor 1000 convertible and in April 1958 I went along to Cowley and collected her personally: this was still possible in those days and maybe still is, but I doubt it. I chose this car as being small enough for the average mule to pull out of a bog or ditch, not too dear in case it became a total loss, for its excellent steering and road-holding, and its fairly large engine in a small and well-tried chassis. In fact, I still consider this model to be Issigonis' best effort of many remarkable designs and I parted with £515 for it quite cheerfully.

This journey passed off uneventfully, but very enjoyably and I described it in some detail in *Autocar* of 15 August, 1958 under the title "To Asia Minor by Morris Minor". About the only thing which fell off were the hood irons, which worked loose on approaching the Turkish border after traversing some of Yugoslavia's rougher roads. I retightened these, which meant stripping out some of the upholstery and trim in the full heat of the midday sun, no shade being available in those parts, and subsequently fitted Simmonds Stop Nu to the bolts which hold them to the bod

The roads in the south of Yugoslavi were distinctly primitive, as indeed som still are, but quite passable in the summ months and the only damage in the who run was a dent in the sump when the c was craned on to an Italian boat on the wa home at Piraeus en route for Bari throug the Corinth Canal. At the end of this run some 4,800 miles the car was so good th I decided to keep her, and I have her to th day.

At 100,000 miles the engine, st original, became rather noisy and the c pressure, when hot, was down to 40lb, s I dropped the sump for the first time ar fitted new big-end shells, but left th mains alone and did not even draw th pistons. This enabled me to run on to total mileage of 130,000 when Numb two plug oiled up and the engine lacke compression on this cylinder. So I remove the head and found the two middle exhau valves in a bad way: moreover, and f more serious, number two piston crow was wet and partly broken away whe parts of the top ring had come o upwards. I turned the engine over with sinking heart and found, as I feared, th cylinder wall was scored. Naturally th depressed me a lot, though I felt that th car did not owe me much at this advance mileage. However, it did look as if she ha reached the end of her road.

Left: The car as it is now after covering 150,000 miles. The bodywork is surprisingly good although the hood (inset picture) is in none too good a state and there is some rust around the joints of wings and body

Right: The engine itself still looks clean and is original, but there is evidence of the high mileage covered in the general underbonnet appearance

Below: Alterations to the interior have included the fitting of flashing indicators and extra instruments, and modification of the passenger-side glove locker to enable it to be locked. There is a Smith recirculatory heater fitted just below the facia

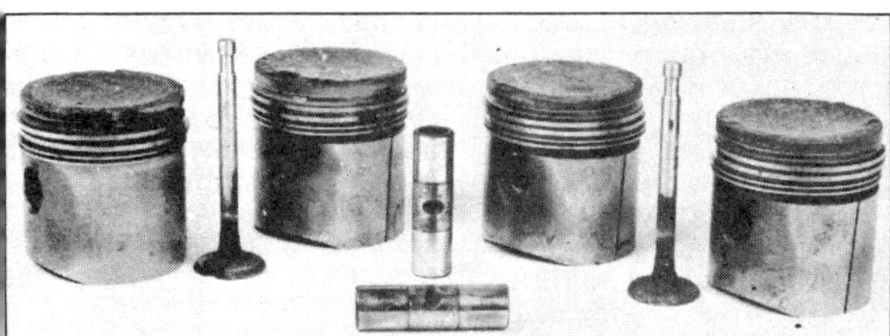

Above: These are the original pistons and two exhaust valves which were replaced after 130,000 miles. Note the partly broken crown of number two piston (extreme left)

Left: Supplementary Smiths water temperature and oil pressure gauges have been added on the left and right of the speedometer respectively. Oil pressure on fast tickover is still a good 40 lb

I sat down on a garden chair in my garage/workshop and took stock of the situation. The car was 12 years old with a big mileage behind her and if I were to get a reconditioned engine, there would still be all those imponderables like the starter, dynamo, and gearbox. The market value of the car, even with all these dealt with, would still be just about nil, and it didn't seem worth it. On the other hand anything in the way of a suitable replacement for shopping and the like would probably set me back around £500, so I decided to take a chance.

I have no reboring equipment, so I looked out some emery cloth and paraffin, with which I rubbed down the worst edges of the score on the damaged bore, and went out and bought a set of four standard size pistons. With a head gasket and so on the total expenditure worked out at just under £10.

I drew the other three pistons, which also had broken top rings, put in the new Hepolites and decoked the head. When all was assembled, I pulled the starter wire with some trepidation, but the engine burst into life almost instantaneously and with only a slight tap. When warm, this wore off and to my amazement the engine used no oil whatever! I had reckoned that if I got 10,000 miles out of my £10 overhaul I should be doing quite well. I have by now covered nearly 20,000 miles and am using only just a trace of oil between changes.

So far there has only been one involuntary stop on the road, which was due to a broken clutch toggle mounting. Borg and Beck declined all responsibility for this, although the pressure plate had only been in use for some 5,000 miles when it happened, and they fell in my estimation for this. The only other part of the car which has been at all troublesome is the two middle exhaust valves, which burnt out at varying intervals. I have tried endless different valves at widely different prices and from the following table it would seem that the quality of valve steel is not to be relied upon, specially as in her early days the car was cruised at 58 mph as against 52–53 mph nowadays.

Below: The Morris Minor outside the Blue Mosque in Istanbul on the original trip to Turkey in 1958

Mileage between removals of head	Type of valve removed
25,470	As assembled at works
19,702	No record
16,890	,,
15,100	,,
6,150	,,
9,245	Standard valves ex BMC
22,196	Special valves ex W. G. James
14,998	Special valves ex Alexander Engineering

Known affectionately as "Nellie" the car has been all over Europe from Oslo to Istanbul, from Finland to the Algarve. The main secret, if it can be called one, of her long life lies in regular and frequent oil changes. I use Castrol and used to change the engine oil every 1,500 miles, which has now become every 2,000; the filter element is replaced every 6,000 miles.

The gearbox, not usually the most reliable part of these cars, has never been touched apart from nine oil changes.

Long may it continue! □

Buying Secondhand

Morris Minor 1000

IN THIS REVIEW of a secondhand car, we are taking a look at a real "Golden Oldie". Britain's Morris Minor was first shown in 1948, and remained in strong, though declining, production until 1971. Until overtaken by the later Minis and 1100s, it was BMC's best-ever selling car, with more than 1½-million sold.

It has very wrongly been called the "British VW Beetle" many times, if only because it remained in production for so long. Whereas the VW was persistently, even rather feverishly developed in its final years, the Minor had its final mechanical updating in 1962, and ran for eight more years without change. Whatever its merits when introduced, or especially when upgraded to become the Minor 1000, by the 1960s it had become thoroughly out of date. It was out-moded, that is, except for a splendid reputation for reliability, and for quite generous four-seat accommodation.

There hasn't been a truly *simple* British Leyland car like the Minor 1000, though its spirit lives on in the Marinas. The Minor was one of those excellent models which almost every garage knew well, and had a lot of space around the engine to allow for service and repair.

Attitudes to the Minor changed greatly over the years. At first, by comparison with pre-war designs, it was thought to be an underpowered little sports saloon. As the Minor 1000, with more power and more potential, it had success in rallies and races. Later still, technical developments and fashion overtook it, and it became known as the district nurse's car.

But this is somewhat of a cautionary tale. While there are Minors on the market in good health and with much potential life ahead of them, there is hardly a single one less than five years old. Rust has become a problem, and many body and coachwork spares are becoming scarce or high-priced. Prices reflect age, and it follows that there is a great deal of ageing competition around. The Morris Minor's contemporaries changed dramatically over the years – in the 1960s one would be looking at Standard Pennants and Triumph Heralds, early Vauxhall Vivas, Austin A35s and A40s, Ford Anglias, Prefects and even an early Escort.

It is quite impracticable to review all versions of the Minor, so we limit our attention to the Minor 1000, first seen in October 1956, re-engined to 1,098 c.c. in October 1962, and running through to the first months of 1971.

How the Minor 1000 developed

The original Minor was designed and developed by Alec Issigonis' team at Morris Motors, Cowley, directed by Sir Miles Thomas. It was intended to have a flat-four engine, but inherited the old Series E side-valve unit instead; this explains the width of the engine bay. During the years a complete range of two-door saloon, four-door saloon, tourer (a convertible) and Traveller (estate) emerged, each in basic or de Luxe form. The car had a unit-construction body (still advanced engineering in the 1940s when conceived), torsion bar independent front suspension, but a conventional leaf-sprung rear axle. Passenger accommodation was quite generous (the roof line was nearly 5ft off the ground), the price was always competitive, and the car soon made itself a reputation for good roadholding and response.

After four years the Austin A30's A-series engine was fitted along with lower gearing and a dreadful wide-ratio gearbox. After another four years, thank goodness, the Minor 1000 appeared, with an enlarged, more robust and more powerful 948 c.c. engine, higher and closer-ratio gearing, and a remote control gearchange. For the first time the car could achieve 70 mph. It needed better brakes (a fault never rectified – Minors never had discs), and was always sparsely trimmed and fitted out. Recognition points over the obsolete Series II machine were the one-piece curved windscreen, and the wrap-around rear-window glass.

The full range of body styles persisted, and there were eight models. The tourer was unique with its fixed hood rails over the doors and rear quarter glass, while the Traveller (a really usefully shaped estate) had BMC's own recognizable type of composite wood and metal construction. Incidentally, while the saloons and tourers were built at Cowley, with bodies provided from Pressed Steel across the road from the assembly lines, Traveller bodies were made by Morris Bodies in Quinton Road, Coventry. Between 1960 and 1964, Travellers were finally assembled alongside MGs and Austin-Healeys at Abingdon.

In addition to private cars, there were, of course, the vans and pick-ups without which no farmer or Post Office fleet would have been happy.

BMC continued to churn out Minors for their clamouring customers. By 1960 production was up to 3,000 every week, and the millionth example rolled off the line in January 1961. But even then there was no doubt that time, and technical progress, were catching up with the Minor 1000. Austin's Farina-styled A40 was a very useful competitor, and when the new front-wheel-drive 1100s, with their Hydrolastic suspension, were revealed, it looked like the end of the Minor.

Not so. The car was given the same 1,098 c.c. engine (it had a longer stroke and a bigger bore dimension than before), the same type of baulk-ring gearbox synchromesh, raised gearing, and proceeded to sell as well as ever! Of course, the car should have been named a Morris 1100 but it never was, and we doubt if many customers even noticed the change.

From October 1962 to early 1971 the cars continued without important changes. When British Leyland was formed in 1968 the new bosses discovered that the cars were actually being sold at a slight loss, and that the production line was taking up a great deal of space. When Cowley was gutted to make way for Marinas, the Minor production lines had to go. There were a few similarities in the Marina's

The Morris Minor as originally introduced in 1948 with side-valve engine. The basic structure remained unchanged for over 20 years

The Minor four-door saloon as it was after the 1962 revisions. Note incorporation of flashing indicators in rear lamp clusters

Probably the best-loved of all the Minors was the convertible, although the folded hood never did look very tidy

)ne of the best-known shapes on the roads of Britain. The side and indicator lamps are combined under one cover

'he "half-timbering" on the rear of the Traveller did support the panels, and any rot can involve a lot of replacement work. The rear floor was quite igh, with the spare wheel, tank and tools located beneath

Buying Secondhand

suspension, and even common parts at first, but nothing more.

The tourer, a low-demand variant, was dropped in 1969, and the saloons followed at the end of 1970. Travellers, with their Coventry-made bodies, continued for a few months, but by April 1971 the last one had been assembled.

Right to the end, many panels, components, even major sub-assemblies, could be interchanged with the parts as designed for 1948. Even if an owner cannot find new spare parts, he can certainly go along to his local scrap yard and find a good selection of suitable machinery. It is also worth remembering that engine parts were common with several other BMC models (the BMC 1100 for instance and the A40) while the A35s and A40s shared the transmission and axle gearing.

What to look for

Remember, first of all, that there have really been two completely different species of Minor 1000. The first cars, built from 1956 to 1962, had the original 948 c.c. engine, constant-load synchromesh gearbox and a 4·55-to-1 axle ratio. From October 1962 the cars were fitted with the long-stroke Morris 1100 type of 1,098 c.c. engine, a gearbox with better baulk-ring synchromesh, different ratios, and a tougher clutch. The axle ratio had been raised to 4·3 to 1.

Quite clearly the post-1962 Minor 1000s are the most desirable – if for no better reason than that all others are now at least 13 years old. Apart from the cars' mechanical condition when inspected, we can see little point in choosing later rather than earlier Minor 1000 "1100s". Mechanically the models were unchanged for eight years, and only minor legislative requirements were satisfied by change.

What about spares availability? In the case of the Minor 1000 there is one perfect example. Last year Dutton-Forshaw (British Leyland distributors) of Swindon, were commissioned to build a brand-new Minor 1000 – from spare parts! The customer could not be dissuaded, even when given an estimated cost of £4,000. Dutton-Forshaw found all but literally one or two tiny items (mainly trim components) from their own BLMC, or other garage spares resources, and delivered the only 1974 Minor 1000 in the world for an actual cost (including assembly labour) of £3,600!

This proves that Minor spares exist (some, in fact, are still being made as new replacement parts), and that their prices are high compared with a manufacturer's OE costs – but there is nothing strange in that.

Dutton-Forshaw's experiences, as related to us by director/general manager Percy Gerrish and service manager Vic Taylor, tell us much about the spares situation. Much was available off-the-shelf, but prices had risen dramatically since the 1960s. Trim items (door panelling, carpets, and the like) were difficult to find. Sheet metal parts – even a basic welded-up body shell – were easily found. Most big Morris (BLMC) outlets still carry wings, bonnets, and doors.

The facia is ultra-plain, with just a speedometer and fuel gauge in the centre, with switches beneath

Despite its short length, the Minor provided reasonable accommodation for four; there was a two-door option

There is a huge amount of room round the engine; the large duct on the left is the heater air intake

Like the Traveller, the spare wheel and tools are stowed under the boot floor. The lid has to be propped open

Front and rear wings, incidentally, are bolted (not welded) to the main structure – a great blessing in case of minor accident damage.

Tourers, it seems, are now rarely seen on the market, and body spares unique to this model are difficult to find. Estates, on the other hand, are popular, and command high prices. Note, however, that the ash items forming the basis of the rear ends are no longer made. Rot and even woodworm are not to be taken as jokes – protection and maintenance is vital to keep this model going; the wood is structural, and cannot be thrown away when past redemption.

Rust in the structure usually shows up around rear spring front mounts – worse on pre-1962 models. Repair, by plating, is possible, and may be demanded when MoT inspection takes place. Front chassis members rust through in time, underneath the car, and might need plating. Front wings, particularly near the front door line and around headlamps, are prone to rusting; rear wings in the usual places (behind wheels, under wheelarches and where vulnerable to road filth) also suffer, and the very bottom of doors and side panels go "frilly" after lengthy use.

Mechanically, the engines have a very good reputation (70,000 miles before piston rings/valves/bearings need replacement is common). The uprated "1100" clutch and the baulk-ring gearboxes are the most desirable items; they can be fitted in place of older components if available. In the same way the later "1100" engine slots easily into earlier 1000 cars with some attention to detail, wires, pipes and suchlike. On the other hand, a Marina 1,275 c.c. unit does not fit without a struggle, as the gearbox and its mating faces are completely different.

We have all seen the broken down Minor 1000 with one front wheel adrift and tucked under the suspension. This is always caused by the threaded joint between kingpin and lower wishbone coming adrift after bad wear and thoroughly bad maintenance. This sort of thing *should* have been picked up in MoT inspections. Front dampers, which also comprise top wishbones, may at times work loose and may also soften off. Check for both items before buying. Axles and axle shafts last for ever if not neglected, but replacements are not now available on the reconditioned scheme.

Unlike Mini and 1100 steering systems, the Minor's rack-and-pinion lasts virtually for ever if kept greased. Brakes, on the other hand, might suffer from seized wheel cylinders. The effect is immediately obvious – efficiency suffers and the "pedal" seems very hard. Reconditioning will usually restore the workings at no great expense.

Do not be put off by exuberantly resonant exhaust systems, for which the Minor 1000 was renowned; do not, however, expect them to last very long either. Replacement, at least, is very simple, as is almost any mechanical adjustment and repair on this simple little car.

Note, finally, that although passenger space is acceptable, the boot is none-too-large. Because of the swept tail, the shape is awkward and large luggage may be difficult to stow. □

Milestones

October 1948: Morris Minor originally announced. First versions had a four-cylinder, side-valve 918 c.c. engine, and low-mounted headlamps. Basic monocoque structure, torsion bar front suspension and spritely roadholding were to remain unchanged for more than 20 years. (Headlamps were raised to their familiar wing-high position in 1950.).

October 1952: BMC A-Series ohv engine of 800 c.c. replaced the old side-valve engine. Lower gearing and wider ratios than before. The Minor Traveller was introduced months later, in spring 1953. Car known as Series II.

October 1956: Minor 1000 replaced the Series II. Basically similar but with 948 c.c. engine, 35 bhp (net) at 4,800 rpm, higher axle (4·55 to 1) and closer gearbox ratios. One piece curved screen in place of vee-screen, and much larger rear window. The fuel tank was enlarged from 5 gallons to 6·5 gallons the following May. Range comprised two-door and four-door saloons, two-door Traveller and a two-door Tourer.

October 1961: Minor changes including addition of seat belt anchorages, screen washers (on de Luxe models), flashing indicators in place of semaphores, and engine improvements. Glove box lids were *deleted.*

October 1962: 1,098 c.c. A-Series engine replaced the 948 c.c. unit, with power output up to 48 bhp. Gearbox was given baulk-ring synchromesh (top three ratios only), instead of the original constant-load arrangement; second gear ratio was raised at the same time, and the final drive ratio became 4·3 to 1.

September 1964: Starter/ignition switch combined. Glove box lids re-introduced. Two-spoke steering wheel. Improved fresh air heater specified. No mechanical changes.

There were no further important mechanical improvements, though the car continued in production until 1971. The Tourer was discontinued in June 1969, saloons in December 1970, and the Traveller, finally, in April 1971.

Spares prices

Component or sub-assembly	Minor 1000 2-door	Minor 1000 4-door	Minor 1000 Tourer	Minor 1000 Traveller
Engine assembly, 948 c.c. (exchange)	£98.28	£98.28	£98.28	£98.28
Engine assembly, 1098 c.c. (exchange)	£98.28	£98.28	£98.28	£98.28
Gearbox (constant load synchro) (exchange)	£57.24	£57.24	£57.24	£57.24
Gearbox (baulk ring synchro) (exchange)	£57.24	£57.24	£57.24	£57.24
Clutch assembly	£14.87	£14.87	£14.87	£14.87
Crown wheel and pinion	£38.61	£38.61	£38.61	£38.61
Axle half shaft	£22.68	£22.68	£22.68	£22.68
Exhaust system complete	£6.70	£6.70	£6.70	£6.70
Front brake shoes (set of 4)	£10.58	£10.58	£10.58	£10.58
Rear brake shoes (set of 4)	£9.42	£9.42	£9.42	£9.42
Front suspension lever-action damper (each)	£17.86	£17.86	£17.86	£17.86
Rear suspension damper	£13.79	£13.79	£13.79	£13.79
Front suspension pivot bushes (set)	£0.08	£0.08	£0.08	£0.08
Dynamo (exchange)	£17.80	£17.80	£17.80	£17.80
Starter motor (exchange)	£11.15	£11.15	£11.15	£11.15
Front door shell (complete)	£28.08	£27.27	£28.08	£28.08
Bonnet assembly	£33.48	£33.48	£33.48	£33.48
Front wing panel	£24.30	£24.30	£24.30	£24.30
Windscreen (toughened)	£9.07	£9.07	£9.07	£9.07

Performance data

	2-door saloon 948 c.c.	4-door saloon 948 c.c.	Traveller estate 948 c.c.	4-door saloon 1,098 c.c.
Road Tested in *Autocar* **of:**	24 June 1960	14 Dec. 1956	27 June 1958	8 May 1964
Mean Maximum Speed (mph)	73	73	69	73
Acceleration (sec)				
0–30 mph	7·2	6·8	7·1	6·6
0–40 mph	12·0	—	14·7	9·9
0–50 mph	18·8	18·8	19·3	16·1
0–60 mph	32·6	31·3	34·1	24·8
0–70 mph	—	—	—	—
Standing ¼-mile (sec)	23·5	24·2	23·9	22·8
Top gear (sec)				
10–30 mph	—	14·9	—	—
20–40 mph	14·6	15·3	15·6	13·6
30–50 mph	17·9	18·2	19·3	16·3
40–60 mph	26·1	23·9	21·6	21·9
50–70 mph	—	—	—	—
Overall fuel consumption (mpg)	34·7	39·4	38·0	31·2
Typical fuel consumption (mpg)	37·0	42·0	42·0	35·0
Dimensions				
Length	12ft 4in., 12ft 5·5in. with overriders			
Width	5ft 1in.			
Height	4ft 10in.			
Weight (lb)	1,785	1,764	1,827	1,708

Approximate selling prices

Price Range	2-door de Luxe	4-door de Luxe	2-door Tourer	Traveller Estate
£100–£150	1963, 1964 and some 1965 models			
£150–£200	1966	1965	1966	1965
£200–£250	1967	1966	1967	1966
£250–£300		1967	1968	
£300–£350	1968	1968	1969	1967
£350–£400	1969	1969		1968
£400–£450	1970			
£450–£500		1970		1969
£500–£550				1970
£600–£650				1971

Note: The price of a new four-door Minor rose from £445 (basic) in 1956 to £451 (basic) with the introduction of the 1,098 c.c. engine, and finally to £592 at the end of 1970 when the saloons were withdrawn.

There has never been a transmission option (overdrive or automatic) on the Minor 1000, and every one was built on cross-ply tyres.

Chassis identification

	Model	Series	Chassis No.
October 1956: Minor 1000 first produced. Original series and opening chassis numbers:	Two-door	FD	448801
	Four-door	FA	448801
	Tourer	FC	448801
	Traveller	FL	448881
October 1962: Minor 1000 with 1,098 c.c. engine introduced, from:	Two-door	M/A2S5	990292
	Four-door	M/AS5	990290
	Tourer	M/AT5	990292
	Traveller	M/AW5	990290
January 1967: Chassis numbers had reached the following points:	Two-door		1170439
	Four-door		1170445
	Tourer		1170452
	Traveller		1170441
June 1969: Convertible discontinued. Final chassis number:			1254328
December 1970: Saloons discontinued. Final chassis numbers:	Two-door		1288377
	Four-door		1288299
April 1971: Traveller discontinued. The last production Morris Minor chassis number was therefore:			1294082

Classic Choice

MORRIS MINOR

Britain's most popular classic? *Paul Skilleter* looks at the Minor market-place and tells you what you get for your money.

ONE of the best-known and most significant of British small cars, the Morris Minor made its bow at the 1948 Motor Show. It was very much the product of one man, Alec Issigonis, and the rightness of its design, its practicality and reliability, and its undating, characterful looks ensured that its career lasted until 1971. Saloon, tourer, Traveller, van and pick-up body styles featured over the years, but here we're going to concentrate more on the saloon versions, in both two and four door modes.

The Minor is virtually unique amongst older cars in that it's still sought after by what one might term 'non-enthusiasts', who value it for normal, every-day motoring. So if you do buy a Minor, you'll find it very practical to live with and drive, even in modern traffic conditions — which can't be said of every car from the 'fifties and 'sixties. Even the earliest examples don't feel too old-fashioned from behind the wheel, although you should think carefully about how you're going to use your Minor before you decide which type to get.

This is because there were three main phases of Minor, beginning with the original Series MM side-valve car of 1948, powered by the Morris Eight 918cc engine; then following the merger with Austin, the Minor was given the A30 overhead valve engine of 803cc, at first in the four-door saloon (from October 1952) and then four months later, in the two-door and tourer models; these were collectively known as Series 11 Minors. Very early Minors, up to October 1950 in the case of the four-door saloon, also *looked* different, as they carried their headlamps in the radiator grille. After this date the lights were moved up onto the wings, and other models followed suit in January 1951, all because of North American lighting regulations and much to the disgust of Issigonis, who has hated the added-on headlamp pods ever since! The final major change occurred in October 1956 when the car was dubbed the Morris '1000' and given an enlarged, 948cc, version of the 'A' series engine; other significant changes included the substitution of a full-width windscreen for the famous 'split' one, and revised interior controls and trim. Last big mechanical revision came in 1963 with the 1098cc engine, which upped performance considerably.

So there's your choice — if you want the most historically significant Minor and don't anticipate covering large mileages, then you might like to search out a Series MM, the low headlamp model being the rarest. Or you may prefer the convenience of 'A' series engine motoring, in which case the split-screen Series 11 might be for you — though don't expect it to be much faster than the MM, which did about 64 mph.

But for sheer practicality, it's the '1000' which is hard to beat; its performance and equipment are more suited to today's motoring conditions (though semaphore indicators lingered on until the end of 1961!), the parts situation is even easier (it's very good for all Minors), and you won't loose out on value — market trends show plainly that a really good 1000 is worth just as much (if not more) than even a very early MM.

Bodywork

There are very few 20-year-old cars where you can go to the manufacturer and order new body panels — and get them. BL Parts can still supply such vital items as front and rear wings, and make the astonishing total of 600 a week of these for Minors. This means that in the case of 1000s at least (Series 11 rear wings, and of course early Series MM front wings, differ) it is feasible to consider buying a rusty car and rebuilding it; but it still means trouble and expense so it's best to find the most unrusted car you can. Here are the points to watch out for.

Starting with the front wings, rust comes through from mud trapped around the indicator light (where fitted) and the headlamp bowl; then a ludicrous design feature at the rear of the front wings (which were never changed in this respect) ensures that unless precautions are taken, the wing will invariably rust from top to bottom adjacent to the door. Under the bonnet the engine compartment valances rust at the bottom, and adjoining the wing at the top, but this isn't too serious.

On opening a front door you'll see a removable step-plate or outer sill — the condition of this again isn't too serious but look underneath the car to see how badly the main sill itself is affected. While you're on your hands and knees, carefully examine the cross-member on which the front suspension torsion bars anchor. If badly gone, the floor under the front seats will also have rotted in a line across the car, which is bad news indeed.

Inside the car, lift the carpets and look all round the edges of both front and rear floor pans for holes. Rot also sets in at the bottom of the outer side quarter-panel (on two-door cars), the effects being visible from outside. Real danger points concern the rear spring mountings front and rear, and it's best to poke these areas with a screw driver if the owner will let you (or isn't looking).

Rust here often extends into the wheel arch, especially where this lies adjacent to the perimeter-type frame which stiffens the Minor's monocoque. The outer rear wings suffer too, round the edges. Open the boot and look at the condition of the floor in front of the spare wheel, particularly in the corners. The bootlid itself rots and at the moment there's a scarcity of new ones. The door construction on a Minor is surprisingly complicated, and rather encourages rust along the bottom — they also crack at the top under the quarter-light on two-door cars. Offside doors are no longer obtainable new, so fight shy of a car with really bad ones.

Most of the chrome trim on the 1000 is still purchasable from your local Austin Morris dealer, except for the bonnet 'mascot', and so are the lights including the big plated rear ones — albeit at a prohibitive cost (something like £27 each). Chrome parts for earlier cars can be difficult, at least brand new. Leather or plastic upholstery and trim was used, and a good interior is worth a lot, though with specialists such as the Minor Centre at Bath about to reproduce such items as the 1000's plastic door panels (they already supply headlinings and seat kits), a tatty interior is not irrecoverable (pun intended).

Mechanics

To include suspension under this heading, everyone's familiar with the Morris parked by the roadside with a front wheel tucked at a peculiar angle under the wheelarch. This collapse occurs when the swivel fails on the front suspension upright, almost invariably because the owner (or previous owners) haven't greased it. Unfortunately it is not always readily apparent on driving the car that it's about to happen, but clues are a lack of grease around the swivel, and, if you can jack the front up, a good degree of 'drop' as weight is taken off the wheel. Otherwise the front suspension is sturdy and effective, contributing greatly to the car's good handling characteristics, though if worn out completely is expensive to rebuild (about £70 per side if you replace everything, and that doesn't include labour). At the rear, we've already covered spring mountings, and the only other component likely to cause slight problems is the lever-type shock absorber, which like the similar front ones, doesn't remain effective for ever. The superb rack-and-pinion steering should be responsive and free from any free play.

Properly maintained, the venerable 'A' series engine almost goes on indefinitely, and with a precautionary change of bearings every 50,000 or 70,000 miles, can go 300,000 miles without a rebore. Its other advantage is the universal availability of spare parts, either over the counter new or from your local scrapyard. The easiest way to overhaul a Minor is simply to go out and buy a 'new' engine from a reputable breakers (or hear it running) for around £35 and slip it in — a mere

Above, Traveller version was always popular in the countryside; woodwork has to be good for MOT. Below, standard car for the district nurse, the 948 Minor initially retained semaphore indicators

day's work. The same applies to the gearbox, which if noisy or tends to jump out of gear, can be dealt with in a similar way. Recon. boxes are obtainable everywhere, from £50 to £80.

Signs of ill-health from the engine are a lack of performance and bearing rumble (or rattle), though the two are not necessarily combined — a rattly engine can perform jolly well (less internal friction?) whereas sluggish acceleration usually points to cylinder head trouble, probably a burnt valve. As everything can be rebuilt or replaced (at moderate cost by today's standards), the condition of a Minor's bodywork is much more important than its mechanical state of health.

Brakes are hydraulic of course, and should work well; about the only point to note is the inconvenient placing of the master cylinder under the driver's floor, where it is awkward to top-up, and a bit of a fiddle to replace — going by the book, one should remove the suspension torsion bar to get at it, though thousands have merely bent the bar for access to the bolts, and lived.

Prices

These vary markedly, the range being roughly from £10 to £4,000; it depends what you want, and who you are. Those buying a Minor instead of a new car for every-day motoring usually go to a dealer, pay £1,200-£3,000, and get a year's guarantee and (according to the price paid) an as-new car; at least from the aforementioned Minor Centre. Enthusiasts, on the other hand, are best advised to purchase their local paper, Exchange & Mart, and T&CC, and peruse the classifieds; to be honest, the first mentioned is probably the best way to buy a reasonable Minor cheaply, the most popular range for a 1000 being from £200 to £600. Tourers start at £300 or so, and go much higher, and so do Travellers though they'll be the subject of another Classic Choice because they're a bit special.

The secret is not to be in a hurry, and while you musn't expect to buy a 30,000 mile one-owner car for £300 (though it has been done), you should end up with an MoT'd saloon for less than £150, and a well-looked-after if not rust-free specimen for no more than £450. For the impecunious do-it-yourselfer, the Minor represents fresh hope and a real opportunity — there's still plenty around "for spares or repair" for £10-£25, and a home rebuild is not an impossible prospect as it is (or should be) with many cars.

The purchaser of the Series 11 and (especially) the Series MM Minor will almost certainly be the 'pure' enthusiast; while there's nothing you can't get to keep a sidevalver going, for instance, maintenance requires a little more perseverance and quite frankly, the MM is nowhere near as practical as the 1000; perhaps it is a car where its historical appeal is now greater than its merits as transport. But whatever Minor you buy, I'll guarantee you'll be captivated by its vivaciousness and will marvel at the brilliance of a concept which is still valid thirty-one years later.●

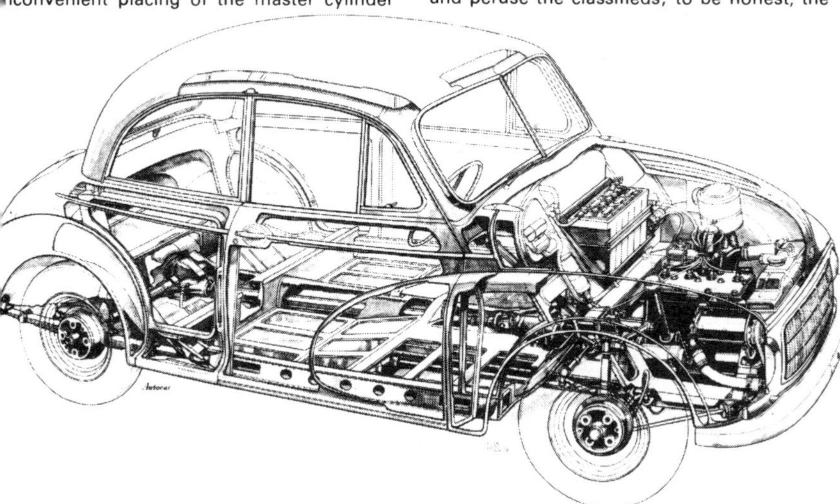

First of the line, the Minor MM with 918 cc side-valve, had headlights inset into the grille but the rest of the design, bar split-window, lived on

Minor front wings trap mud front and rear — this wing has certainly gone beyond repair, although obviously someone's had a go!

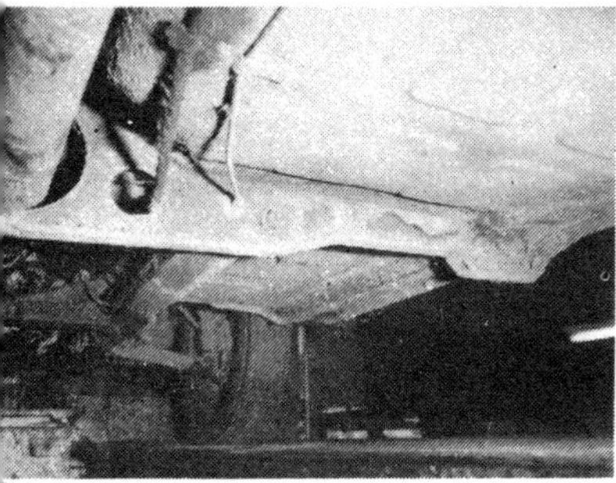

ront suspension torsion bars anchor on this ross-member which also carried the jacking oints. It usually rusts towards its outer nds, and where it meets the floor.

Sills carry drainholes but still rot; note too the hole in the outer skin of the rear side panel, and the front mounting of the rear spring, around which rust can get a hold.
Right: floors can go as badly as this
Below: typical engine compartment of a 1098 cc Minor — lots of space to work in, notice! Fuel consumption rarely rises above 35mpg, and not much restraint gets 40mpg or more, which is another Minor virtue. Late Minors will cruise happily at 65-70mph, and approach 75mph.

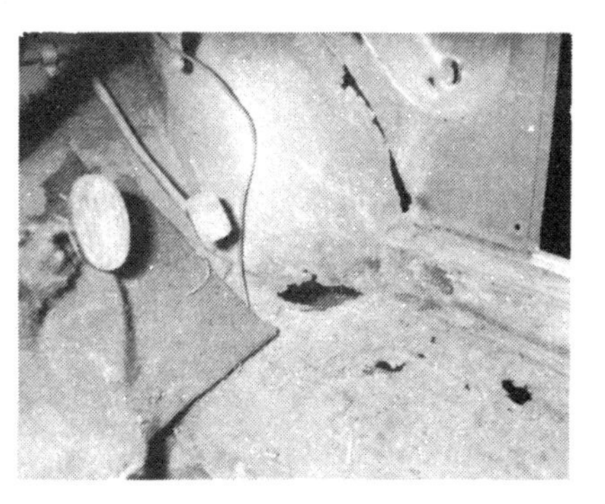

An advanced case of rear-end rot, extending from the rear spring shackle and through the inner wing into the boot, below

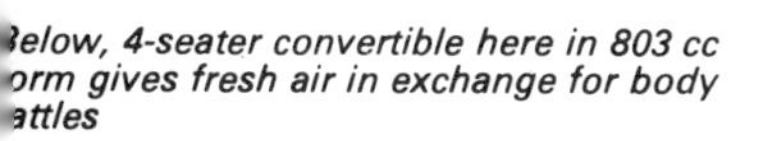

elow, 4-seater convertible here in 803 cc orm gives fresh air in exchange for body attles

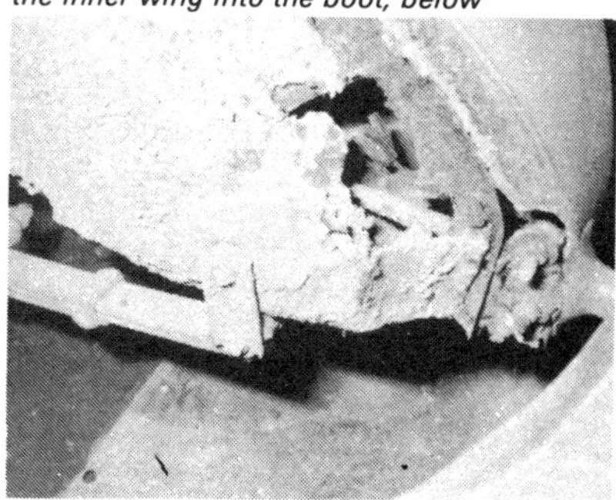

POWERPLUS MINOR . . .

CONTINUED FROM PAGE 45
the clutch plate torsion springs seems desirable. Moving the gear lever out of second needed more effort than is normal, but the newness of this particular car may have been the cause. Fast downward changes require double declutching, for the synchromesh is easily beaten.

Braking was safe for the increased performance, but on occasions there was a feeling that a little more stopping power would have been appreciated. Adjustment to recover pedal travel was necessary after 1,000 miles of mainly hard driving.

On indifferent road surfaces, the Minor's enhanced performance was curtailed by the energetic behaviour of the rear axle, which gave the occupants a harsh ride, particularly when the car was lightly laden. The urge to make a rapid getaway from rest is likely to be felt by owners of Minors with this conversion. If take-off is too violent, axle hop develops and hard acceleration before straightening up from a slow corner sometimes produces a form of axle tramp. Steering tremor, to which rack and pinion layouts are prone, begins at about 70 to 75 m.p.h. depending upon surface conditions, but causes no anxiety.

Possessed, then, of a performance comparable with many saloons of twice its engine capacity, the Powerplus Minor 1000 also shows outstanding economy. Driven consistently near the limit of its performance, a fuel consumption of 38 m.p.g. was recorded over 247 miles. Heavy city traffic reduced this to 33 m.p.g. Normal though spirited driving without exceeding 60 m.p.h. in top gave a creditable 47.6 m.p.g. There was no pinking on premium petrol.

Malaysia: Morris 1000 (secondhand)

AA British influence in Malaysia may have declined since this Morris Minor 1000 was delivered new in 1961 but the model's popularity is still unchallenged. It holds its price well in a flourishing market for British secondhand cars because it meets the demand for basic motoring that is simple, orthodox and easy to maintain. The same applies in Britain.

The inside feels high and narrow with plenty of headroom, even hat room. You sit unusually upright at the wheel which comes close to the driver's chest, a position which can be extremely tiring on a long journey, and the gear lever and handbrake are also close to the driver.

Four people have reasonable elbow room and good legroom but the boot is wedge-shaped and you need either wedge-shaped luggage or squashy bags. Rear doors are rather narrow.

When it first appeared the Minor was well ahead of its time in handling and roadholding. It is still acceptable; there is only a little body lean on corners and it feels very sure footed. Although the steering has enough feel to help the driver keep control easily, accelerating out of corners can make the back axle hop.

The engine is smooth and revs freely. It gives a game rather than vigorous performance, and you have to change gear quite often on hills to maintain speed. A fuel consumption of 35 to 40mpg should be within reach of most drivers.

The windscreen is narrow and the pillars unfashionably stout. The windscreen wipers leave large areas unswept.

The test car showed that Minors can stand up to seven years' use well. The transmission was noisy and the gearbox synchromesh was worn so you had to change gear slowly to avoid clashing sounds. Likewise the clutch had to be let in gently to avoid a lurch on starting. The ride was good and the shock absorbers still worked well. The heater was feeble, even with the booster fan on full.

Kuala Lumpur

Ong Thye Keow, 50-year-old sawmiller, transports his thirteen children by Morris Minor 1000—a fleet of three, new and secondhand.

I bought my first Minor in 1955 and it has become part of the family. It is reliable and dependable and I shall keep it as long as it will run.

The conservative shape attracted me. Cars that don't change their shape keep their value. My latest Minor was expensive to buy for its size and age but it gives 40mpg, roadholding is steady and it is easy to manoeuvre. The manufacturers should strengthen the shock absorbers, because our roads weaken them, although none of the cars has needed many repairs. The cost of maintenance is very low.

SPARE PARTS

	UK	Malaysia
New engine	£62	£90 12
Gearbox	35 10	56 1
Clutch	7 5	4 17
Silencer	1 8	1 14
Tyre	6 6	6 2

PRICE	UK	Malaysia
now	**£220**	**£435**
new (1961)	£679	£840 15

COST		
per week	**£3 14s**	**£6 13**
running	3 0	2 15
depreciation	14	3 18

SPECIFICATION
Engine—948cc, 4 cylinders, 37bhp at 5,000rpm, oil change every 3,000 miles.
4 forward gears, no synchromesh on first.
Suspension—front independent, rear leaf.
Brakes—drum all round. Tyres—5 x 14.
Fuel—6½gal. Body—4-door saloon.
Colours: black, grey, white, green, blue.

The Specialists

Minor Shrine

Paul Skilleter visits the Morris Minor Centre at Bath

THE Morris Minor enjoys a virtually unique position in the old-car world, because what other obsolete model are people deliberately turning to in preference to a brand new car for every-day motoring? That's actually happening, thanks to the Minor's innate reliability, practicality, and economy — especially in recent months when 40-plus mpg from a non-depreciating asset suddenly seems to make extra good sense.

No-one is more aware of the Minor phenomenon than Charlie Ware, as he sits in his busy little office overlooking a yardful of twenty or thirty of the ilk in various stages of undress. Charlie started buying and selling old cars just like hundreds of other small-time motor traders up and down the country, except that few of them had previously made (and lost) a million pounds from buying and restoring Georgian buildings. At first he dabbled in anything — Anglias, Minis and so on — but then the Minor began to exert its peculiar appeal; he found them simple, solid, and with lots of room inside — and he found that people liked them.

Charlie decided to specialise, and founded the Morris Minor Centre in and around one of the slightly run-down Georgian houses he used to repair, so that he could sell or rebuild cars to or for the faithful.

But his customers weren't — and aren't — what you and I would label "enthusiasts"; no, they're largely family motorists, often into their forties, to whom the Minor is a personal friend to be cherished for its own virtues and for its practical value as sensible workaday transport — not as a mere weekend toy. Disturbed by the increasing reluctance of their local garage to stock parts or patch up the bodywork, the arrival of the Minor Centre in 1977 appeared as a ray of hope. Someone cared after all, and didn't laugh.

Charlie was lucky — or successful — with his publicity. The *Sunday Times* feature spread the news of the Minor shrine at Bath like wildfire, and that was followed by a generous 15 minutes on "Pebble Mill" which alone produced 5,000 letters. It was initially a little more than the fledgling workshops could manage, but then techniques were developed to cope, and an idea gained of what people really wanted.

So the Minor Centre has a variety of cars for sale, ranging from sound to almost-like-new, with prices starting from £900 and going up to £4500 for an utterly rebuild Traveller; yes of course you can buy a Minor for less than that, although Charlie doesn't sell them direct from the Centre itself.

Oh yes, Charlie guarantees his cars, and reckons to have sold a thousand Minors since he started getting involved with them. His warranty claims must be the envy of any car manufacturer, being reckoned at no more than 1 or 1½%; he did have one customer come back recently, after twenty months' everyday motoring — the speedo had packed up. But then the Minor always did have one of the best warranty-claim records of any British car when it was sold new. The sensible and simple specification explains why, and when it does need to be repaired, it can virtually be unbolted and then put back together again in a weekend. The bodywork is only marginally less susceptible than average to the tin worm however, though Charlie has plans for an all-new galvanised body.

Customer's cars are taken in for rebuilding, although only to a strictly set standard and it's not a particularly cheap process, the bill rarely being below £1500 unless the car is already exceptional. But then any job well done is hardly ever cheap, and the money's not wasted in the long term, thanks to the appreciation in value of the type. All sorts undergo the treatment, from a (very few) sidevalvers through convertibles to four- and two-door saloons and the most expensive (to buy and repair) Travellers.

Then there's parts, and Charlie is putting a lot of effort into this side of the business. A very comprehensive, profusely illustrated catalogue is being drawn up and parts sources vigorously researched and extended. Thanks to the popularity of the car a surprising amount is still available from BL (though often the local dealers will brush-off would-be customers with stories to the contrary), and the situation is going to get better as the Minor Centre commissions replica parts as items go out of stock at BL. Eventually, additional branches in key parts of the country will be established, simply to retail Minor spares.

An offshoot of the Centre's fame are the regular approaches from owners with low-mileage or well-kept Minors who've given up driving or through other reasons want to dispose of their car. They contact Bath because they've heard that the people there also care about Minors, and will find a good home for theirs which must, with reluctance, be sold. They know then that it won't be sold to Belgium for soup, as Charlie puts it, referring to less fortunate four-legged friends. But Charlie is careful never to "steal" a car, and usually pays a lot more than the owner would obtain from their local garage or advert in the local paper.

All the money goes straight back into the business; a lot of building has been going on in what originally looked like a mid-sixties one-make scrapyard behind Charlie's establishment, and now there's an efficient body shop, garage and spray booth. The recent, abitrary, rises in the price of fuel has noticeably increased interest in the Morris Minor and the Centre is busier than ever; after all, 40 mpg is obtainable with very little restraint, and 50 mpg is within reach under the right conditions. One could also envisage a special economy version of the Minor, perhaps the 948cc ohv engine with the Riley 1.5 differential with its high 3.7 ratio.

There's thought to be something around 200,000 Morris Minors still on the road in this country alone, which is the reason why Charlie is investing so heavily in the future. Along with owners, he is refusing to acknowledge that the car is obsolete and will, I think, help create an influential lobby which will encourage BL to maintain production of essential parts. This can only help everyone. If you want to contact the Minor Centre, it's on the Lower Bristol Road, Bath, Avon, and the telephone number is 0225 315449. ●

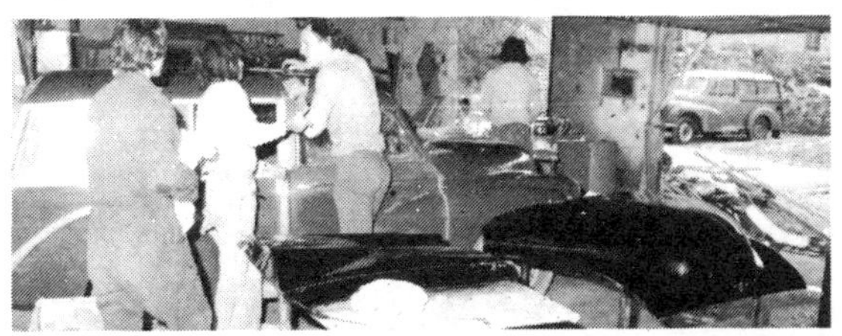

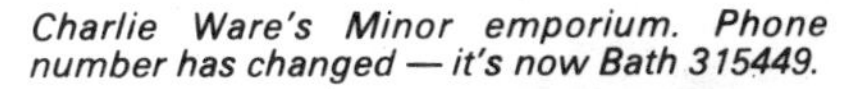

Charlie Ware's Minor emporium. Phone number has changed — it's now Bath 315449.

Door and bonnet being cut back after spraying. A sea of mainly "A" series engines.

Thou good and faithful servant

CONTINUED FROM PAGE 57

is a bit sloppy in the housing of the original motor. It's getting worse so I check monthly. One day . . . . The 1957 coil was replaced by a Lucas HA12 high-output one with some mild tuning in prospect. It enabled the plug gap to be opened out to 25-thou. and the S.U. nut weakened by seven flats. I can't claim that this has improved the starting for the engine has always started immediately, but it has improved fuel consumption. Tuning has prevented a performance comparison with the standard Minor 1000. Oil consumption is about 500 miles per pint, bore wear just detectable, compression excellent and the fuel usage remains at 30 m.p.g. in London, 34 town and country, 38-40 on holiday, and I cruise at a speedo 65 m.p.h.

The Minor 1000 has a handle, of course, and it starts readily on that too; but as the valve clearances *never* need resetting, and the car always starts even though it's left out, the handle is relegated to working the dreadful B.M.C. jack, something which comes up more often than I like because the rear brakes devour lower shoes every 5,000 miles. The conical corrector spring that B.M.C. supply does not stop the shoe canting over. Other shoes had been wearing at the rate of 40,000 miles I suppose, but I hadn't got through a set of fronts when a brake mod was put in hand. I wanted more stopping power with less pedal pressure, especially at the bottom ends of long gradients so this is currently being achieved by fitting a set of Wolseley 1500 front brakes plus a Lockheed servo and a set of Ferodo VG95 competition linings.

I aligned the fan-dynamo-crankshaft pullies recently, serviced the heater motor, replaced the flexible brake hoses, freed a jammed window winder, put in another tailpipe, renewed all the rubber bushes in the front suspension (which is by torsion bars), had another go at draught-sealing the back doors, and dealt with a collapsed parcel shelf by fitting the wider one belonging to the two-door variant. The heater tap disintegrated and had to be replaced, too.

In a wild fit of enthusiasm recently, I blanked off the disused semaphore-signal openings with stainless-steel strip and re-stuffed the driver's seat. Youngsters occur in most families, a fact of which too few car designers seem to be aware, for much should be done to make things safer for them. A set of Wilmot-Breeden childproof doorhandles were well worth the price in peace of mind.

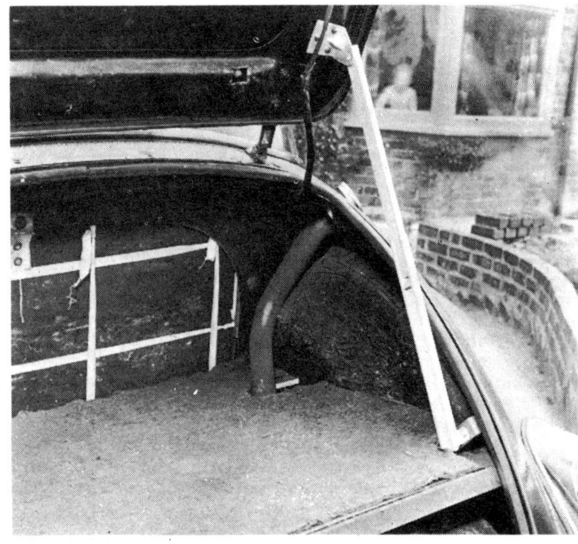

Boot lid stay is from a modern Minor and replaces the old hand-placed strut. If evenly distributed, a load of over 3 cwt. can be carried without the headlamps going noticeably high.

The body is remarkably free from corrosion and this small damage along a door sill was the worst part that could be found on the whole car.

What attention is the car likely to need soon? That's a touch-wood-quick sort of question. I would say new synchro cones; I doubt if the clutch will outlast teaching my wife to drive but she hates the idea of going to a school; one of the replacement i.f.s. rubbers went recently and the others look like following it very shortly; there's starter trouble round the corner and I suppose that one day the white-of-egg plus Holt's Rad-Seal will wash out of the radiator. Neither the prop-shaft nor the Armstrong dampers are going to last for *ever*, surely?

Logically, I ought to expect the need for new swivel pins, steering rack adjustment and rubbers and ball joints, steering column felt bushes, new back axle parts, wheel bearings, handbrake cables, petrol gauge, distributor, a.v.c., dynamo, stop-lamp switch, and radiator. I know that the wiper motor, the stone-chipped windscreen, the rust on the front bumper, corrosion where the left rear wing abuts on to the body, a broken boot hinge, fractured bonnet cross-member, worn brake and clutch pedal bushes and some gear lever chatter are going to catch up with me. Until then, they can wait. Undersealed from new, and rust-inhibitor-painted by me, the body and chassis have shown little deterioration.

The only job which I felt was a waste of time was renewal of the front and rear engine rubbers, but new ones in the head steady were worth-while. The alarming discovery was made of irremediable metal-to-metal contact of the gearbox U-brackets with the chassis cross-member. Inspection of other Minors shows that mine is not alone in this.

I agree with Alf Long that black is about the last colour to choose. Is this enough to cause me to overcome my inertia and respray the car? The accident damage has left a piebald effect in several different shades of black. Or will I buy a new car? And if I did—the same again? I need more luggage space and the dog's a pest. A Traveller version? High on my list of priorities is a marque and a model for which the vendor maintains a 100 per cent spares service. I am choked with shopping around several B.M.C. agents for Morris spares, and being sent elsewhere for Lucas, Smiths and Lockheed items. Only Stewart and Ardern of Acton, and Wadhams of Guildford meet this criterion in the territories I frequent, and neither is conveniently near. I dislike vehicles with engine and transmission in one lump as being too cumbersome for private-owner repairs. I don't like those with low door sills and don't particularly want to buy larger or foreign. I want sophisticated ventilation and a starting handle.

Since I'm aiming for another 50,000-ish from my Morris, there's plenty of time for some more choice to arrive on the market. For reliability, economy of operation, finish and general specification I'd go B.M.C. again.

M

MILES Behind the Wheel

Mighty Minor

It has 300 bhp, does 0-100 mph in 9.5 sec, and takes some beating up the hills

VAY FROM the high-flown anced circus of motor racing re is still the world of hill- nbing and sprints. There the ented amateur can even now to grips with the man who afford to buy success; haps not so much in the tright classes but certainly ver down the scale. This is ere brains, resourcefulness, d in Nick Man's case, a bit of very behind the wheel, can l shame the monied man.

During the day Mann works for o Research in Abingdon. After urs, he nurtures what has to be fastest accelerating Morris nor in the world – a car that rently harries and sometimes ats much more sophisticated chinery up the hills. It's by far fastest road legal car to 100 h tested by *Autocar*. On the d it is shattering.

As an amateur Mann has done uperbly professional job. He ows exactly what he is doing d why he is doing it (all too e). You cannot build a car like s on a salary unless you do.

Two years ago, after some rs of steady development ng a turbocharged MGB gine (latterly inter- oled, and blowing at some psi), Nick decided that the time s right to try and get the car y competitive up the hills, the viso being that it should nain street-legal at all costs. st important this, because rt from the clutter and ense of a suitable tow car and ler Mann pointed out (on eral occasions) that "One of best parts is driving it a ple of hundred miles across ntry to a meeting early in the rning."

began to understand what he ant as we headed out from his ntage home across country vards Chipping Norton and a able aerodrome for testing rby. I gently opened the ottle to overtake a truck. The gine soon cleared of its low- vn carburation richness, nged to a gruff growl and . . . ack. It was like being smacked he back by something heavy soft – a slowly swinging chbag perhaps. The truck was the problem, the sudden ase of adrenalin was! eleration was so vicious that I to concentrate very hard – e eyed – on the gap I was ng for if I was not to overshoot

What on earth was it going to like on full throttle or when rging off the line? How was he ng to convert it from a road into a racer? Answers come r.

rst, the car. Almost elievably the centre and rear y sections are quite dard. In a rather elegant e of fabrication (remember the whole car was built in a lock up garage) most of the front bulkhead has been cut away to allow the Rover V8 to sit well back. A space frame is attached to the lower chassis rails and the now suitably strengthened firewall/bulkhead to carry the engine, front suspension, radiator (MBG), and glass fibre nose section. Unequal length wishbones are nicely fabricated in square tube. Vauxhall Ventora uprights, discs and calipers are used. After much soul-searching Nick purchased some *brand new* Spax dampers with adjustable spring abutments.

Most surprising is that a Mk2 Cortina rear axle (acquired like many other parts from a breaker's yard) has stood the strain for two seasons without any trouble; though Nick admits "the halfshafts twist a little so I replace them from time to time." The limited slip unit is a clutch type set up tightly. Longitudinal location of the axle is by four equal length links, and lateral location by a long Panhard rod. Rear suspension is by Spax coil spring damper units as at the front.

Mann's experience with turbocharging gained through the ealier MGB turbo engine proved most useful in successfully converting the Rover unit with the diverse set of ancilliaries (some of which he has had to make from scratch) he had at his disposal.

The Rover bottom end was rebuilt using "a certain proprietary piston" (he would not say which one) to lower the compression ratio to 6.0 to 1. A Crane solid lifter kit replaced the standard hydraulic tappets. The heads were left virtually untouched and retain standard valves. When it came to the turbo installation Mann was out on his own. Instead of going for a conventional suck or blow through carburettor set up, he opted to try and adapt Tecalamit fuel injection. In order to obtain the required on-boost enrichment in an all-mechanical system that relies simply on throttle position and engine revs, he added a boost-enrichment device sensing from both sides of the butterfly (made from an SU dashpot) to command the necessary extra fuel flow.

On the exhaust side, the sheer lack of space prevented sophisticated manifolding. Each bank dumps exhaust into a pipe set at right angles and within a few inches of the ports. The manifolding runs back to a large Switzer turbocharger supplied by commercial vehicle turbocharger reconditioning experts BTN Turbocharger services–Mann's only sponsor, and then mainly in kind with wheels and tyres etc. The butterfly is immediately upstream of the compressor. Air is forced into the engine down a single 2in pipe connecting to an experimental Tecalamit manifold, in which the injectors are sited, as close as possible to the port faces. Boost is wastegate-controlled at 15psi for the road. For the hills (and our shattering runs) Mann removes the air cleaner pipe (with scant regard for the compressor vanes) and puts about eight turns more

Left: Bulkhead is cut away to allow the Rover V8 to sit well back, and the engine squeezes in with only inches to spare. Tecalemit injection involves much pipery. Inside, engine cover lifts to reveal the turbocharger and wastegate (right). Boost adjustment is via the screw on top – a simple matter from the driver's seat

Steering column is lengthened, rally type seats are mounted well back and angled to point occupants away from the engine cover. Dash contains (from right) speedometer, dial containing ammeter, fuel and water temperature gauges, revcounter, large oil pressure warning light and boost gauge

Above: Mann with his Minor; entire front end is glassfibre moulding

download on his home-brewed wastegate. Then he can cut loose with the full 22psi.

It was typical of the man to ignore "expert" advice not to use the SD1 gearbox. He acquired a "reject" unit which has since given no trouble, possibly because during hill climbs the potentially weak fifth gear is prevented from engaging by an outside locking mechanism which is only switched out for road use or if you need much more than 100mph in a speed event.

A typical morning of an event will require Mann simply to bolt on the SP-shod tin wheels, load up his four slick-shod Compomotive rims, the bolt-on wheel spats, the scissor jack, fuel can, his sandwiches, thermos, helmet, overalls, and a pair of old Adidas training shoes, then set off, which is more or less what we did.

Double wishbones are fabricated in square tube and attach directly to the front space frame

Morris Minor V8 Turbo

True mph	Time (secs)	mph	Top	3rd	2nd
30	1.9	10- 30	–	–	–
40	2.5	20- 40	–	–	2.2
50	3.1	30- 50	–	5.1	1.7
60	4.4	40- 60	5.6	2.4	–
70	5.3	50- 70	4.7	1.8	–
80	6.6	60- 80	3.1	2.2	–
90	8.0	70- 90	2.6	–	–
100	9.5	80-100	3.0	–	–
110	12.5			1949	

Standing ¼-mile: 12.8sec, 113mph

Aluminium boxing surrounding the engine intrudes well into the driving compartment. Because of this you sit quite a long way back, and with your legs threading into the narrow footwell. It looks uncomfortable but isn't. The door trim is still there, wind up windows, wipe/wash indicators, etc. The gearlever is cranked forward about a foot to be within sensible reach, which gives you some idea how far back the engine is. The car incidentally is slightly rear-heavy. Mann quotes weight at 16 cwt, distributed approximately 48/52 front to rear.

We trundled out of Wantage. It seemed a surprisingly couth road car, the only real niggle being that in order to get good full throttle engine response Mann has to run over-rich settings thus low-speed and part-throttle pick up suffer. As the engine began to warm so the low down fluffiness increased. Once nursed through this period it flew. We eased our way out of town. The limited slip made graunching and chattering noises round sharp corners, the engine note was nothing more than a subdued waffle. Apart from the noisiness induced by solid-bushed suspension the car rode poor surfaces surprisingly well; better still on the open road. But this was a racer, so why such road-car-like suspension settings? "Hills" explained Mann, "are invariably bumpy, narrow, and have mostly short corners. You need traction, you need a car that will ride the bumps and give good bite and feel into the corners. It doesn't have to be stable through long fast corners. We run road specification pads and linings, and very soft compound rubber because you need instant braking response – instant grip – at relatively low speeds and for short periods."

It rides the bumps well. Directional stability is adequate, if not arrow-like, which is fair enough with 300bhp and say 350lb ft of torque in a short wheelbased softly sprung car, on road tyres. But the poke – that is simply unbelievable. You would move out, have a look, then point the car, and then and only then squeeze the accelerator. As hell lets loose one has to watch the tendency to wander. The boost gauge flashes round. There is a hard growl. The whole car seems to contort. Simultaneously your back tenses to take the thrust, there is no time to look at the rev counter dashing round to the 6,000rpm limit – the power curve is flat-topped anyway. Can you believe 50-70 in third takes 1.8 sec? That's nearly a second faster than an Aston Vantage. That 70-90 in fourth takes 2.6 sec against 4.2 sec in third for the Aston? – Figures that were recorded *before* we turned up the boost and changed to the racing tyres, and on the sort of burning hot day which hardly assists in reducing underbonnet temperatures, nor might I add the temperature of the occupants. We were sweating, believe me!

The serious stuff would start after the five minute conversion to competition specification. The slicks stand outside the body but are much smaller diameter so do not foul. The spats were quickly attached with 2BA bolts and nuts. We were still on 15psi boost. Starts are made in the 60 mph second gear. Taking the car off the line is that delicate balance between no wheelspin at all (and the engine going off turbo), or setting light to the rear tyres. Nick blips the throttle to clear the bottom end fluffiness and feeds in the clutch – doesn't simply drop it – but feeds it in quickly to cushion the driveline from shock.

Head down, looking at the fifth wheel speedometer unwinding, one is conscious even above the combination of induction noise, blower whine and the rumbling growl, that the rear wheels have broken traction with a screech: The front is rearing like a frightened horse. A good start means the wheelspin and tyre vibration fade as we pass through 30mph, Nick smacks the lever through from second to third with a crunch. The front falls and rises in sympathy. The car continues its headlong rush There's just time to pull out fifth gear lock, before takin fourth, then at 100mph the the dog-leg change to fifth.

We did three runs with th 100 mph time hovering aro 10 sec. We even managed sec on the road tyres – this a muffed start.

Up came the inside engir cover, out came a man-size screwdriver to tune in anot 7psi boost. Off the line one the extra edge. There are n mistakes or muffed change With an initial shudder the picks itself up. At 30mph th watch stops at 1.9 sec, 4.4 at 60 mph and a stunning 9.5 at 100mph, then on to 110 in sec. It was all over.

I could add little during r play. The starts are criticial over eargerness has to be controlled. Although pretty heavy, the clutch is positive bites well (though Borg and thought it might not). This m accurate starts more a matte getting the engine revs righ Then it's a question of forcing gearchanges through against SD1 box's notchiness, and remembering to move out fifth gear lock (a plunger on side of the transmission tun at the appropriate time. The rise rapidly but power tails o the danger of over-revving engine is small. You want change up all the time.

On the drive back discuss was of future developments interposed with the occasio dart past traffic, though sa say all the Oxfordshire Pors Turbo owners seemed to b home that day. Whether th chassis would accept more power is a moot point. Nic agreed that if room could found, intercooling would obviously increase output substiantially. He reckoned among other things, winter developments might include tubos sucking through SU carburettors; or perhaps a different car altogether with intercooled Rover Turbo en behind the driver. We shall

As it is, he has recently top a string of seconds and third Shelsley, Prescott and Lotor Park, with a win and class re at Gurston Down. Not bad f car which eleven years ago being used by his mother f shopping.

It looks deliciously normal when fitted with road wheels and tyres – the ultimate Q-car?